SpringerBriefs in Applied Sciences and Technology

Thermal Engineering and Applied Science

Series Editor
Francis A. Kulacki, University of Minnesota, USA

For further volumes:
http://www.springer.com/series/10305

Seshasai Srinivasan • M. Ziad Saghir

Thermodiffusion in Multicomponent Mixtures

Thermodynamic, Algebraic,
and Neuro-Computing Models

 Springer

Seshasai Srinivasan
Department of Mathematics
    and Statistics
McMaster University
Hamilton, Ontario
Canada

M. Ziad Saghir
Department of Mechanical
    and Industrial Engineering
Ryerson University
Toronto, Ontario
Canada

ISSN 2191-530X   ISSN 2191-5318 (electronic)
ISBN 978-1-4614-5598-1   ISBN 978-1-4614-5599-8 (eBook)
DOI 10.1007/978-1-4614-5599-8
Springer New York Heidelberg Dordrecht London

Library of Congress Control Number: 2012950589

Printed on acid-free paper

Springer is part of Springer Science+Business Media (www.springer.com)

To our families.

Preface

The subtle and intricate phenomenon of thermodiffusion is evident in several facets of life and nature. However, it is still not understood very well and there is a lack of a single source that deals exclusively with this phenomenon. Consequently, a newcomer in the field of thermodiffusion has to fend for himself through the research articles that are scattered all over the engineering and pure science journals. In an effort to address this shortfall, this text is an attempt to offer a comprehensive starting point by presenting the current theoretical/computational approaches to study thermodiffusion.

In writing this book it has been assumed that there is little or no prior understanding of the principles of thermodiffusion. In laying down the foundations of thermodiffusion, the book discusses the current theoretical frameworks to study this fascinating phenomenon. Additionally, studies employing these theories are presented for the reader to understand the drawbacks/advantages of the various expressions that quantify thermodiffusion. A separate list of references is also included at the end of each chapter for further reading.

Apart from serving as a reference text for the professionals in the fields of chemistry, mechanical engineering, reservoir engineering as well as materials science, this book is also an excellent guide for the postdoctoral researchers and graduate students in this field.

In preparing this work, we are indebted to our fellow researchers in the scientific community whose studies have helped us shape our understanding of the subject. Special thanks to the members of our research group, graduate students in particular, for their contributions to this fascinating field.

Hamilton, Canada   Seshasai Srinivasan
Toronto, Canada   M. Ziad Saghir

Contents

Acronyms

BLGA	β-lactoglobulin $-A$
CFD	Computational fluid dynamics
DM	β-dodecyl maltoside
IBB	Isobutylbenzene
ISS	International Space Station
LNET	Linear nonequilibrium thermodynamics
MEK	Methyl ethyl ketone
MN	Methylnaphthalene
NaPSS	Sodium polystyrene sulfonate
PC-SAFT	Perturbed chain statistical associating fluid theory
PMMA	poly(methyl methacrylate)
PR	Peng–Robinson
PS	Polystyrene
RMS	Root mean square
SDS	Sodium dodecyl sulfate
THF	Tetrahydrofuran
THN	Tetralin
v-PR	Volume translated Peng–Robinson

Chapter 1
Introduction

Abstract *Thermodiffusion* is a coupled mass and heat transfer phenomenon in which the constituents in a mixture develop a preferential separation direction in the presence of a thermal gradient. It occurs in numerous areas in nature and is also employed as an underlying principle in the separation processes in engineering fields. In space platforms like the International Space Station, this phenomenon has shown some promise as a mode of transport of fluids. This chapter will introduce the phenomenon of thermodiffusion as it is studied in the various mixtures in different disciplines. Application and significance of thermodiffusion are also presented to outline the motivation behind the thermodiffusion studies. We also briefly highlight the challenges in some of the studies that are conducted in outer space. The chapter concludes with an outline of the rest of the book.

1.1 Background: Definition and History

By applying a nonuniform temperature distribution to a mixture, one can observe a transport of its constituents. This phenomenon was first observed by Ludwig [23]. However, nearly two decades later, Charles Soret independently discovered this phenomenon and subsequently studied it in greater detail [41]. In recognition of these two researchers, *Thermodiffusion* is also known as the *Ludwig-Soret effect* or simply the *Soret effect*. Thus, a formal definition of this phenomenon can be stated as:

> *Thermodiffusion* or the *Soret effect* is the separation of the components of a mixture toward the hot/cold regions of a domain that is characterized by a nonuniform temperature distribution.

S. Srinivasan and M.Z. Saghir, *Thermodiffusion in Multicomponent Mixtures*,
SpringerBriefs in Applied Sciences and Technology, DOI 10.1007/978-1-4614-5599-8_1,
© Springer Science+Business Media, LLC 2013

Acknowledging the contributions of Charles Soret in discovering and understanding this phenomenon, it is pertinent to discuss his initial works in this phenomenon. In 1979, he published his pioneering findings in the field of thermodiffusion by investigating the effect of nonuniform temperature on salt solutions [41]. Specifically, the objective of his investigation was to understand if the homogeneous concentration in the NaCl and KNO_3 solutions were retained even in a nonuniform temperature field. The experiments were conducted in two different types of tubes that contained these solutions, the ends of which were maintained at different temperatures. His conclusion was that salt concentrated at the cold end. He also studied KCL [42], LiCl [42], and $CuSO_4$ [43] solutions, and arrived at the same conclusion. In his works he also quelled any speculative thoughts about the thermodiffusive separation being an artifact, by suspending the solutions in air without any thermal gradient, and demonstrating that the concentration of the solution at the top and bottom was almost the same.

1.2   Relevance of Thermodiffusion

Thermodiffusion has been found to influence several natural activities such as convection in stars [44] and the physics of a solar pond [51]. In the analysis of the distribution of components in oil reservoirs, the thermodiffusion effect coupled with isothermal and pressure diffusion are considered important [27]. A salinity gradient in ocean caused by temperature difference is due to the Soret effect and is called Thermohaline convection [37].

Thermodiffusion is the underlying principle in the separation and characterization of polymers [15, 20, 21, 36]. The field-flow-fractionation devices are based on this phenomenon [6, 24–26, 35, 48], and are used to separate bacteria [16, 17], colloidal materials [14, 50], charged particles [29], etc. Another application of thermodiffusion includes isotope separation in liquid and gaseous mixtures [11]. The Soret effect is also speculated to play a role in the freezing process of foods [30].

The Soret effect can also contribute in some biological applications. Bonner and Sundelöf [4] suggested the use of thermodiffusion as a mechanism of biological transport. Biological applications such as determination of biomolecular binding curves [1] and trapping of DNA [5] make use of the principles of thermodiffusion.

1.3   Studies in Thermodiffusion

Such a wide sphere of influence of thermodiffusion has prompted extensive research, computational as well as experimental, in various types of mixtures, viz., gases, electrolytes, liquid hydrocarbons, alcohols, ferrofluids, polymers, proteins, molten metals, semiconductor materials, surfactant micelles, and latex particles. Studies of this phenomenon on such a large variety of mixtures has resulted in this phenomenon being known via alternative names such as *Thermophoresis*, *Thermotransport*, and *Thermomigration*.

The experimental research has led to the development of several methods to study thermodiffusion, including the classical Soret cell, thermogravitational column, the two-chamber thermodiffusion cell, thermal field-flow fractionation, the microfluidic fluorescence method, laser-beam deflection technique, thermal diffusion forced Rayleigh scattering, and the thermal lens technique. An initial review of the experimental approaches was presented by Lin and Taylor [22]. More recent review articles on the experimental approaches to study thermodiffusion have been presented by Platten [32], and Srinivasan and Saghir [45].

Theoretical investigations have resulted in the development of numerous formulations for explaining thermodiffusion in different types of mixtures. These include the models by Drickamer and coworkers [7, 33, 49], Kempers [18, 19], Semenov and Shimpf [38–40], Bielenberg and Brenner [2], Morozov [28], and Eslamian and Saghir [9, 10].

Computational works include the study of thermodiffusion via the principles of molecular dynamics as done by Galliero et al. [12, 13]. Computational fluid dynamics (CFD) studies, in which complete Navier–Stokes equations, coupled with the thermodiffusion models, are also used to study the thermodiffusive flows. Finally, in a novel approach, artificial neural networks have recently been proposed to study thermodiffusion [46, 47].

Some review articles on the theoretical studies on thermodiffusion are presented by Wiegand [52], Piazza and Parola [31], Blums [3], and Eslamian and Saghir [8]. Summarizing the theoretical models, simulations, and relevant experiments, Wiegand [52] outlined the properties and mechanisms involved in the Soret effect in polymer liquids. On the other hand, Piazza and Parola [31] reviewed the experimental and theoretical methods pertaining to the study of thermodiffusion in colloidal mixtures. They also studied relevant results to probe the aspects of colloid solvation forces and emphasized particle manipulation in microfluidics via thermophoresis. Blums [3] reviewed the experiments on ferrocolloids, pointing out that the magnetic field influences the mass diffusion more strongly, as a consequence of which magnetic Soret effect is much stronger than the theoretical predictions. Finally, Eslamian and Saghir [8] critically examined the theoretical models to study thermodiffusion. They also elucidate the role and significance of the heat of transport and the activation energy of viscous flow.

1.4   Founding Principles of Thermodiffusion Formalisms

As stated before, the thermal gradient will induce a mass separation. However, as soon as there is a concentration gradient, the Fickian diffusion process begins to act in the opposite direction to restore the uniformity of the mixture. Consequently, in studying the fluxes of the species in a binary mixture, the net mass flux of the species is written as

$$J = -\rho \left(D \nabla c + D_{\mathrm{T}} c (1 - c) \nabla T \right), \tag{1.1}$$

where D_T is the thermodiffusion coefficient with units $m^2s^{-1}K^{-1}$, D is the molecular diffusion coefficient with units m^2s^{-1}, ρ is the mixture density, c is the mass fraction of the reference species, and T is the temperature in Kelvin. These coefficients give a measure of the molecular diffusion and the thermodiffusion in the mixture. Now, at equilibrium, when the net flux vanishes ($J = 0$), the above equation can be rearranged as

$$\rho D \nabla c = -\rho D_T c (1 - c) \nabla T, \tag{1.2}$$

or more appropriately as

$$\nabla c = -S_T c (1 - c) \nabla T, \tag{1.3}$$

where S_T is the *Soret coefficient* that is

$$S_T = \frac{D_T}{D}, \tag{1.4}$$

with units of K^{-1}. Another closely related dimensionless parameter is the *thermodiffusion factor* that is defined as

$$\alpha_T = T \frac{D_T}{D}. \tag{1.5}$$

Also, sometimes, the *thermal diffusion ratio* is used that is defined as

$$\kappa_T = \alpha_T x_1 x_2, \tag{1.6}$$

x_i being the mole fractions of the two components in a binary mixture. Usually, in the thermodiffusion studies, the reported results for the Soret effect are either the thermodiffusion coefficient (D_T), or the Soret coefficient (S_T) or the thermodiffusion factor (α_T) or the thermal diffusion ratio (κ_T).

It must be noted that sometimes, in the second term on the right side of (1.1), the product $D_T c (1 - c)$ is collectively written as the thermodiffusion coefficient, in which case the equation becomes

$$J = -\rho D \nabla c - \rho \mathscr{D}_T \nabla T, \tag{1.7}$$

where

$$\mathscr{D}_T = D_T c (1 - c). \tag{1.8}$$

As a result, while comparing the models with experimental data one must be careful about the convention used.

The thermodiffusion coefficient can also be derived using the molar flux equation. Such a derivation of D_T has been done in Sect. 2.2 of Chap. 2. By switching between mass and molar frames, one can end up with different measures of the thermodiffusion coefficients in multicomponent mixtures. However, in binary mixtures, the choice of the frame of reference does not have impact on the calculated values of D_T.

1.5   The Subtle Nature of Thermodiffusion: The Undesirable Convective Forces

Thermodiffusion is a very subtle phenomenon that can be quickly destroyed by the slightest disturbance. Specifically, the small concentration gradient in the mixture can be easily removed if there are convective forces that operate in the opposite direction by introducing mixing in the mixtures. This has been clearly demonstrated in several investigations.

This is true even in optical experimental setups that are deemed accurate due to the absence of mechanical disturbances. For instance, in the investigations of thermodiffusion of gold colloidal particles in mixtures, Schaertl and Roos [34] have found that as time progresses, with laser heating, gold particles begin to oscillate. To mitigate this drawback, the authors advocated studying such mixtures in reduced gravity conditions. There have been other works that have promoted using the reduced gravity environments to overcome the undesirable convective forces that can hinder the thermodiffusive separation.

With the advancement of technology, space platforms such as the International Space Station (ISS) and the free flying satellites are now available to scientists to conduct experiments in such reduced gravity conditions. In fact, several of these studies have successfully demonstrated that the low gravity conditions yield different results. Specifically, with respect to thermodiffusion investigations, it has been found that the concentration gradients are larger with the minimized natural convection.

It is also pertinent to bring to light the different set of issues that arise in conducting experiments on such space platforms:

- *Vibrational forces*: Space platforms suffer from constant vibrations due to several factors such as motion of the platforms and operation of on-board devices. On ISS, there are additional contributors to these vibrational forces such as the activity of the crew when they are working, periodic thrusting forces that are essential to uplift the platform, and spacecraft docking. These vibrations, measured by the on-board accelerometers, have accelerations that are as high as $10^{-4}\,ms^{-2}$. Vibrations on ISS in particular are notoriously high as compared to the free flying satellites that do not carry crew. In fact, on such satellites, the quality of the reduced gravity environment improves considerably, with accelerations that are several orders of magnitude smaller. Nevertheless, the presence of such undesirable disturbances has prompted the investigators to study the effect of vibrations on the thermodiffusion experiments. In this, fundamental investigations include studying the impact of simple sinusoidal vibrations of known frequency and amplitude on a fluid enclosed in a cell that is subject to the thermal gradient. Further, in such studies, the imposed vibrations are usually applied in a direction that is orthogonal to the direction of the applied gradient.

- *Expensive setup*: A major hurdle in studying thermodiffusion in reduced gravity conditions is the huge expense that is involved in conducting such studies. Since the entire process is almost completely automated, there are enormous expenditures involved in designing the special hardware, hauling the experimental apparatus to space, etc.
- *Insufficiently trained crew*: Assembly, initiation, and some tasks pertaining to the experiments on ISS are usually done by the on-board crew. Since these astronauts are multitasking scientists who are not necessarily experts in the field of thermodiffusion, they usually undergo preflight training to assemble and conduct the experiments. However, despite adequate training, a small error in the setup or operation of the experiment in space can result in unacceptable errors. A need for rectification of the experiment can sometimes be done only on the labs on earth, and so rerunning the experiments mean another space flight, loss of sufficient time, etc. Put differently, rectification is very expensive.
- *Reduced opportunity*: In view of the fact that there is intense competition for physical space, very few scientists get an opportunity to study thermodiffusion experiments in such space platforms. In other words, the facilities might not be easily available for the scientists to conduct their investigations.

In summary, it is safe to say that the promising environments to study purely diffusive phenomenon come at a very high cost and have their own set of challenges. It is also important to state at this time that there are investigators who believe that accurate thermodiffusion measurements in some liquids are possible on ground-based conditions. Nevertheless, given that such platforms are available to researchers, it is appropriate to use them exhaustively before arriving at a particular conclusion.

1.6   Topics of This Book

In this book we discuss three approaches to study thermodiffusion in liquid mixtures, viz., thermodynamic, algebraic, and neurocomputing. Each approach has its place in research pertaining to thermodiffusion. At a fundamental level it is always important to develop an understanding of the underlying physics by not only identifying the various sources/parameters that contribute to thermodiffusion, but also to understand the way in which these sources initiate the inter-particle interactions for this complex process.

On the other hand, simulating a complete multicomponent thermo-convection-diffusion fluid motion using CFD tools coupled with the detailed thermodiffusion models that are formulated on our present understanding of the underlying physics can be very time intensive. In fact, in planning simple preliminary experiments to be conducted in the lab, for instance, it might be of interest to just get an approximate trend of the expected results that is perhaps not very far from the true results. In such cases, it might be easier to evaluate some simple algebraic equations that have been deduced from the experimental analysis of previous investigations.

Finally, from an engineering point of view, if one just needs a reliable predictive ability for a wide variety of liquid mixtures of polymers, molten metals, hydrocarbons, etc., then artificial neural networks serve as an ideal solution. This seemingly random and yet sophisticated tool can predict the thermodiffusion data as well as important trends in the process.

The rest of the book is organized as follows: In Chap. 2 we present the theoretical framework and the relevant equations proposed by different researchers for all three approaches, viz., principles of linear nonequilibrium thermodynamics (LNET), algebraic correlations, and the principles of artificial neural networks. A detailed derivation of the first and the last approach is included in this chapter. For the second method, since it is essentially a by-product of the experimental results, the algebraic formulations that have been proposed in the literature for various types of mixtures are presented.

In Chaps. 3 and 4, the application of the LNET models to various types of mixtures from the literature is considered. Specifically, Chap. 3 deals with the application of LNET models to quantitatively estimate the magnitude of the thermodiffusion process in nonassociating, associating, polymer as well as molten metal mixtures for a given temperature, pressure, and mixture composition. The impact of the *equation of state*, that is necessary to calculate the state parameters, is also discussed briefly.

A more detailed scrutiny of the transient separation behavior in a domain containing a fluid mixture is presented via the CFD tools in Chap. 4. In this chapter, case studies are presented for the separation behavior of the components in a ternary and a binary mixture.

In Chap. 5, several algebraic correlations outlined in Chap. 2 are employed to quantify the thermodiffusion process in binary as well as ternary mixtures. Their predictive capabilities are understood by comparing them with the experimental data of the corresponding studies.

Finally, in Chap. 6, the principles of artificial neural networks are applied to study thermodiffusion process in binary n-alkane mixtures and binary molten metal mixtures. To highlight the quality of performance of this approach, it is compared with the other two methods to study thermodiffusion.

References

1. Baaske P, Wienken CJ, Reineck P, Duhr S, Braun D (2010) Optical thermophoresis for quantifying the buffer dependence of aptamer binding. Bioanal Chem 49:2238–2241
2. Bielenberg JR, Brenner H (2005) A Hydrodynamic/Brownian motion model of thermal diffusion in liquids. Phys Stat Mech Appl 356:279–293

3. Blums E (2004) New problems of particle transfer in ferrocolloids: magnetic soret effect and thermoosmosis. Eur Phys J E 15:271–276
4. Bonner FJ, Sundelöf LO (1984) Thermal diffusion as a mechanism for biological transport. Z Naturforsch C 39(6):656–661
5. Braun D, Libchaber A (2002) Trapping of DNA by thermophoretic depletion and convection. Phys Rev Lett 89(18):188,103
6. Cölfen H, Antonietti M (2000) Field-flow fractionation techniques for polymer and colloid analysis. In: New developments in polymer analytics I. Advances in polymer science, vol 157. Springer, Berlin, pp 67–187
7. Dougherty EL, Drickamer HG (1955) A theory of thermal diffusion in liquids. J Chem Phys 23(5):295
8. Eslamian M, Saghir MZ (2007) A critical review of thermodiffusion models: Role and significance of the heat of transport and the activation energy of viscous flow. J Chem Phys 126: 014,502
9. Eslamian M, Saghir MZ (2010a) Dynamic thermodiffusion theory for ternary liquid mixtures. J Non-Equilib Thermodyn 35:51–73
10. Eslamian M, Saghir MZ (2010b) Nonequilibrium thermodynamic model for soret effect in dilute polymer solutions. Int J Thermophys 32:652–664
11. Fury WH, Jones RC, Onsager L (1939) On the theory of isotope separation by thermal diffusion. Phys Rev 55(11):1083–1095
12. Galliero G, Bugel M, Duguay B, Montel F (2007) Mass effect on thermodiffusion using molecular dynamics. J Non-Equilib Thermodyn 32(3):251–278
13. Galliero G, Srinivasan S, Saghir MZ (2010) Estimation of thermodiffusion in ternary alkane mixtures using molecular dynamics and the irreversible thermodynamic theory. High Temp High Pressur 38:315–328
14. Jančа J (2003) Micro-channel thermal field-flow fractionation: high speed analysis of colloidal particles. J Liq Chrom Relat Technol 26(6):849–869
15. Jančа J (2006) Micro-thermal field-flow fractionation in the analysis of polymers and particles: a review. Int J Polymer Anal Char 11:57–70
16. Jančа J, Kašpárková V, Halabalová V, Šimek L, Ružička J, Barošová E (2007) Micro-thermal field-flow fractionation of bacteria. J Chrom B 852:512–518
17. Kašpárková V, Halabalová V, Šimek L, Ružička J, Jančа J (2007) Separation of bacteria in temperature gradient: micro-thermal focusing field-flow fractionation. J Biochem Biophys Methods 70(10):685–687
18. Kempers LJTM (1989) A thermodynamic theory of the soret effect in a multicomponent liquid. J Chem Phys 90:6541–6548
19. Kempers LJTM (2001) A comprehensive thermodynamic theory of the soret effect in a multicomponent gas, liquid, or solid. J Chem Phys 115:6330–6341
20. Kita R, Wiegand S, Luettmer-Strathmann J (2004) Sign change of the soret coefficient of poly(ethylene oxide) in water/ethanol mixtures observed by thermal diffusion forced rayleigh scattering. J Chem Phys 121(8):3874–3885
21. Köhler W, Rosenauer C, Rossmanith P (1995) Holographic grating study of mass and thermal diffusion of polystyrene/toluene solutions. Int J Thermophys 16:11–21
22. Lin JL, Taylor WL (1988) Thermodynamics of thermal diffusion. Technical Report MLM-3614, United States Department of Energy
23. Ludwig C (1856) Diffusoin zwischen ungleich erwwärmten orten gleich zusammengestzter losungen. Sitz. Ber. Akad. Wiss. Wien Math-Naturw 20:539
24. Martin M, Min B, Moon MH (1997) Interpretation of thermal field-flow fractionation experiments in a tilted channel. J Chromat 788:121–130
25. Martin M, Van Batten C, Hoyos M (2002) Determination of thermodiffusion parameters from thermal field-flow fractionation retention data. In: Köhler W, Wiegand S (eds) Thermal nonequilibrium phenomena in fluid mixtures. Springer, Berlin, pp 250–284
26. Messaud FA, Sanderson RD, Runyon JR, Otte T, Pasch H, Williams SKR (2009) An overview on field-flow fractionation techniques and their applications in the separation and characterization of polymers. Progress Polymer Sci 34:351–368

27. Montel F, Bickert J, Lagisquet A, Galliero G (2007) Initial state of petroleum reservoirs: A comprehensive approach. J Petrol Sci Eng 58(3–4):391–402. http://dx.doi.org/10.1016/j.petrol.2006.03.032

28. Morozov KI (2009) Soret effect in molecular mixtures. Phys Rev E 79:031,204

29. Pasti L, Agnolet S, Dondi F (2007) Thermal field-flow fractionation of charged submicrometer particles in aqueous media. Anal Chem 79:5284–5296

30. Pham QT (2006) Modeling heat and mass transfer in frozen foods: a review. Int J Refrigeration 29:876–888

31. Piazza R, Parola R (2008) Thermophoresis in colloidal suspensions. J Phys Condens Matter 20:153,102

32. Platten JK (2006) The soret effect: a review of recent experimental results. J Appl Mech 73:5–13

33. Rutherford WM, Dougherty EL, Drickamer HG (1954) Theory of thermal diffusion in liquids and the use of pressure to investigate the theory. J Chem Phys 22:1157–1165

34. Schaertl W, Roos C (1999) Convection and thermodiffusion of colloidal gold tracers by laser light scattering. Phys Rev E 60(2):2020–2028

35. Schimpf ME (2000) Thermal field-flow fractionation. In: Schimpf ME, Caldwell KD, Giddings JC (eds) F.F.F. Handbook. Wiley, New York, pp 239–256

36. Schimpf ME, Giddings JC (1987) Characterization of thermal diffusion in polymer solutions by thermal field-flow fractionation: effects of molecular weight and branching. Macromolecules 20:1561–1563

37. Schmitt RW (1994) Double diffusion in oceanography. Ann Rev Fluid Mech 26:255

38. Semenov S, Schimpf M (2004) Thermophoresis of dissolved molecules and polymers: consideration of the temperature-induced macroscopic pressure gradient. Phys Rev E 69(2):011,201

39. Semenov S, Schimpf M (2005) Molecular thermodiffusion (thermophoresis) in liquid mixtures. Phys Rev E 72:041,202

40. Semenov S, Schimpf M (2009) Mass transport thermodynamics in nonisothermal molecular liquid mixtures. Phys Usp 52:1045

41. Soret C (1879) Sur L'état d'équilibre Que Prend Au Point De Vue De Sa Concentration Une Dissolution Saline Primitivement Homohéne Dont Deux Parties Sont Portées A Des Temperaturés Différentes. Arch Sci Phys Nat 2:46–61

42. Soret C (1880) Influence De La Temperaturé Sur La Distribution Des Sels Dans Leurs solutions. Acad Sci Paris 91(5):289–291

43. Soret C (1881) Sur L'état d'équilibre Que Prend Au Point De Vue De Sa Concentration Une Dissolution Saline Primitivement Homohéne Dont Deux Parties Sont Portées A Des Temperaturés Différentes. Ann Chim Phys 22:293–297

44. Spiegel EA (1972) Convection in stars-II: special effects. Ann Rev Astron Astrophys 10:261–304

45. Srinívasan S, Saghir MZ (2011) Experimental appraoches to thermodiffusion – a review. Int J Therm Sci 50:1125–1137

46. Srinivasan S, Saghir MZ (2012a) A neurocomputing model to calculate the thermo-solutal diffusion in liquid hydrocarbon mixtures. Neural Comput Appl DOI: 10.1007/s00521-012-1217-6

47. Srinivasan S, Saghir MZ (2012b) Modeling of thermotransport phenomenon in metal alloys using artificial neural networks. Appl Math Modell DOI:10.1016/j.apm.2012.06.018

48. Thompson GH, Myers MN, Giddings JC (1967) An observation of a field-flow fractionation effect with polystyrene samples. Separat Sci Technol 2:797–800

49. Tichacek LJ, Kmak WS, Drickamer HG (1956) Thermal diffusion in liquids; the effect of nonideality and association. J Phys Chem 60:660–665

50. Van Batten C, Hoyos M, Martin M (1997) Thermal field-flow fractionation of colloidal materials: methylmethacrylate-styrene linear di-block copolymers. Chromatographia 45(1):121–126

51. Weinberger W (1964) The physics of the solar pond. Solar Energy 8:45–56

52. Wiegand S (2004) Thermal diffusion in liquid mixtures and polymer solutions. J Phys Condens Matter 16:R357–R379

Chapter 2
Thermodiffusion Models

Abstract Three approaches to study *thermodiffusion* in binary and multicomponent mixtures are explored in this chapter, viz., the nonequilibrium thermodynamics, algebraic correlations, and artificial neural network. The first method employs the principles of nonequilibrium thermodynamics to explain thermodiffusive separation, by considering the heat and mass fluxes in the mixture as linear functions of forces such as temperature gradient and chemical potential. The second method is based on the observation of relations between the thermodiffusion parameters and parameters such as the mixture composition and pure component/mixture properties. Finally, in artificial neural networks, a data mining of a reasonably large set of experimental data is undertaken and a model is developed that predicts the thermodiffusion data based on the principles of associative thinking. To this end, mathematical functions are integrated in the model to quantify the decision-making process. Expressions corresponding to all three methods are discussed in this chapter.

2.1 Linear Nonequilibrium Thermodynamic Theory

In many irreversible processes, the state variables are functions of space and time. Hence, in studying such systems, it becomes essential to formulate the basic equations in a way so as to refer to quantities at a single point in space. In other words, a local formalism of the fundamental equations is necessary. To this end, a theoretical framework for the macroscopic description of the irreversible processes is provided by the theory of nonequilibrium thermodynamics.

The key aspect of the theory of nonequilibrium thermodynamics is the *entropy* balance for an infinitesimally small volume element. *Entropy* is a thermodynamic property that can be used to determine the energy not available for work in a thermodynamic process. It has a unit of JK^{-1}. The change in entropy in this volume element is due to two reasons, viz.,

S. Srinivasan and M.Z. Saghir, *Thermodiffusion in Multicomponent Mixtures*,
SpringerBriefs in Applied Sciences and Technology, DOI 10.1007/978-1-4614-5599-8_2,
© Springer Science+Business Media, LLC 2013

1. Entropy flux into the element
2. Entropy production inside the volume element

Thus, a transport equation for the change in entropy would be of the form

$$\frac{ds}{dt} = -\mathrm{div}\mathbf{J}_s + \sigma \tag{2.1}$$

$$\sigma \geq 0. \tag{2.2}$$

In the above equation, s is the entropy per unit mass, σ is the entropy production per unit volume per unit time or simply entropy production strength. Finally, $\mathbf{J}_s$ is the total entropy flux per unit area per unit time less the convective contribution in this flux. It must be noted that in the transport equation (2.1), the second term on the right side is the source term. Further, since entropy can only be created, for an entropy balance, this source term is always positive, as indicated by (2.2).

Now, in order to relate the entropy source to the irreversible processes in the system, a complete set of macroscopic governing equations is generally written for a local volume element to account for the transport of mass, species, momentum, and energy. Using these and the Gibbs thermodynamic relation, one can develop a relation between the rate of change of entropy and other parameters, viz., rate of change of internal energy, rate of change of composition, and rate of change of specific volume. In particular, it is assumed that within the small mass element of interest, an equilibrium state exists, *local*, if one may add, although the entire system in itself might not be in a state of equilibrium. For such a small element, along the center of mass, we can write the Gibbs equation as

$$T\frac{ds}{dt} = \frac{du}{dt} + P\frac{dv}{dt} - \sum_{k=1}^{n} \mu_k \frac{dc_k}{dt}, \tag{2.3}$$

where u is the specific internal energy, P is the pressure, v is the specific volume and μ_k is the chemical potential.

2.1.1 Phenomenological Equations

As is customary in any set of governing equations, an equation of state is also included for closure. However, despite this, the set of governing equations including the equation for entropy balance and the equation of state cannot be solved with the initial and boundary conditions. This is due to the presence of the irreversible fluxes that are unknown in this set of equations. To address this issue, a set of *phenomenological equations* are also included in this system of equations. These phenomenological equations relate the unknown irreversible fluxes *linearly* with the thermodynamic forces in the entropy strength, and would be called the *linear phenomenological equations*, i.e.,

$$J_i = \sum_k L_{ik} \Phi_k, \tag{2.4}$$

where J_i is a component of the flux in the direction i, L_{ik} are the phenomenological coefficients and Φ_i is the thermodynamic force in the direction i. It must be noted that the entropy production is related to the fluxes and the forces as

$$\sigma = \sum_k J_k \Phi_k. \tag{2.5}$$

In summary, the entire set of partial differential equations consisting of the governing equations, phenomenological equations, and the equations of state is complete, and can be solved using the initial boundary conditions to study the irreversible process.

Extensive details of the theory of nonequilibrium thermodynamics have been presented in the book of de Groot and Mazur [21] and will not be pursued further here. Instead, in the following section, drawing from these principles of linear of nonequilibrium thermodynamics theory, relevant equations pertinent to the derivations of the formalisms for thermodiffusion processes will be discussed.

2.2   Method 1: Thermodiffusion Models Based on LNET Theory

For a n-component mixture at a temperature T and pressure P, if we assume that the only external force acting on the system is gravity (**g**), then we can write the entropy production strength as

$$\sigma = -\frac{1}{T^2} \mathbf{J}'_q \cdot \nabla T - \frac{1}{T} \sum_{k=1}^{n} \mathbf{J}_k \cdot (\nabla_T \mu_k - \mathbf{g}), \tag{2.6}$$

where $\mathbf{J}_k$ is the molar diffusion flux relative to the molar average velocity of the kth component. The vector $\mathbf{J}'_q$ is the heat flux due to pure conduction heat flow and is given as

$$\mathbf{J}'_q = \mathbf{J}_q - \sum_{k=1}^{n} h_k \mathbf{J}_k. \tag{2.7}$$

In the above equation, $\mathbf{J}_q$ is the total heat flux and h_k is the partial specific enthalpy of the kth component. Thus, the total heat flux consists of two parts, viz., heat flux due to pure conduction, denoted by $\mathbf{J}'_q$, and heat flux due to the diffusion of mass, denoted by the term $\sum_{k=1}^{n} h_k \mathbf{J}_k$.

Another method of expressing $\mathbf{J}'_q$ is to use the concept of net heat of transport [8, 54]. As per this representation,

$$\mathbf{J}'_q = \sum_{k=1}^{n} Q_k^* \mathbf{J}_k, \tag{2.8}$$

where Q_k^* is the net heat of transport of the kth component. More specifically, it is the conductive heat flow of the kth component per particle per mole or per kilogram, required to be absorbed by the local region to keep the temperature of the region constant.

Now, keeping in mind that $\sum_k \mathbf{J}_k = 0$, in a n-component system, there are $n - 1$ independent fluxes and so the entropy production strength in (2.6) can be written as

$$\sigma = -\frac{1}{T^2}\mathbf{J}_q' \cdot \nabla T - \frac{1}{T}\sum_{k=1}^{n-1}\mathbf{J}_k \cdot (\nabla_T(\mu_k - \mu_n)). \tag{2.9}$$

Further, incorporating (2.8) in the above expression, we get

$$\sigma = -\sum_{k=1}^{n-1}\left[(Q_k^* - Q_n^*)\frac{\nabla T}{T^2} + \frac{\nabla_T(\mu_k - \mu_n)}{T}\right] \cdot \mathbf{J}_k. \tag{2.10}$$

At this time, one can write phenomenological equations for the diffusion mass flux in terms of the phenomenological coefficients as

$$\mathbf{J}_i = -\sum_{k=1}^{n-1}L_{ik}\left[(Q_k^* - Q_n^*)\frac{\nabla T}{T} + \nabla_T(\mu_k - \mu_n)\right], \tag{2.11}$$

or slightly more explicitly as

$$\mathbf{J}_i = -\sum_{k=1}^{n-1}L_{ik}\left[(Q_k^* - Q_n^*)\frac{\nabla T}{T} + \sum_{j=1}^{n-1}\frac{\partial \mu_k}{\partial x_j}\nabla x_j\right]. \tag{2.12}$$

An additional way of writing the diffusion flux equations is by using the transport coefficients in conjunction with the gradients of temperature, pressure, and concentration. This will yield

$$\mathbf{J} = -\chi\left(\mathbf{D}_M\nabla \mathbf{x} + \mathbf{D}_T \cdot \nabla T + \mathbf{D}_P \cdot \nabla P\right), \tag{2.13}$$

where $\mathbf{D}_M$, $\mathbf{D}_T$ and $\mathbf{D}_P$ represent the molecular diffusion coefficients, thermodiffusion coefficients, and barodiffusion coefficients, respectively, and χ is the mole density. Further, $\mathbf{D}_M$ is a matrix with elements $\mathbf{D}_M = [D_{ij}]$. On the other hand, $\mathbf{D}_T = (\mathscr{D}_T^{(1)}, \mathscr{D}_T^{(2)}, \ldots, \mathscr{D}_T^{(n-1)})$, $\mathbf{D}_P = (D_P^{(1)}, D_P^{(2)}, \ldots, D_P^{(n-1)})$ and $\nabla \mathbf{x} = (\nabla x_1, \nabla x_2, \ldots, \nabla x_{n-1})$ are vectors. In the absence of any pressure gradient, the above equation reduces to

$$\mathbf{J} = -\chi\left(\mathbf{D}_M\nabla \mathbf{x} + \mathbf{D}_T \cdot \nabla T\right). \tag{2.14}$$

From (2.12) and (2.14) we can deduce the expression for the thermodiffusion coefficient as

$$\mathscr{D}_T^{(i)} = \frac{\sum_{k=1}^{n-1}(Q_k^* - Q_n^*)L_{ik}}{\chi \times T}. \tag{2.15}$$

2.2.1 The Net Heat of Transport

From (2.15) it is clear that for an accurate estimate of the thermodiffusion coefficient, precise evaluation of Q^* is essential. As mentioned earlier, Q^* is the amount of energy which must be absorbed by the region per mole of a component while diffusing out in order to maintain the constancy in the temperature and pressure of the mixture. To this end, we can represent it in terms of the energy needed to detach a molecule from its neighbors and the energy given out when a molecule fills the hole. In this, we can use the viscous energy as an approximation of the energy needed to detach a molecule. For a multicomponent mixture, an expression for Q^* has been proposed by Firoozabadi and coworkers [19] as

$$Q_i^* = \frac{U_i}{\tau_i} + \left(\sum_{j=1}^{n} x_j U_j / \tau_j \right) \frac{V_j}{\sum_{k=1}^{n} x_k V_k}, \tag{2.16}$$

where τ_i is the ratio of the cohesive and viscous energy, of component i. U_i and V_i are the partial molar internal energy and partial molar volume, respectively.

2.2.2 Equation of State

An equation of state is used to determine the state parameters such as density, pressure, fugacity, and enthalpy. Depending upon the type of mixture being studied, different equations of state have been developed in the literature. For instance, the Peng–Robinson (*PR*) equation state [43] is more suited to the hydrocarbon mixtures than water alcohol mixtures. For the latter type of mixtures, the Perturbed Chain Statistical Associating Fluid Theory (*PC-SAFT*) equation of state [22, 23] is more accurate. While there are other choices available for an equation of state (e.g., Cubic Plus Association equation of state), the user must extensively evaluate them with respect to experimental data before selecting them for a particular type of mixture. Two equations of state are now discussed below to highlight the difference between their formulations.

2.2.2.1 Peng–Robinson Equation of State

In the Peng–Robinson equation of state [43], the parameters are expressed in terms the critical properties and the acentric factor. It is applicable to liquid as well as gas phase, and is fairly accurate near the critical point. In particular, it has good prediction capabilities for the compressibility factor and liquid density.

Density: To calculate density, the following equation is used:

$$\rho = \frac{P}{ZRT}M, \tag{2.17}$$

where Z is the unknown compressibility factor, R is the universal gas constant, and M is the mixture molar mass. The pressure, P, is the second unknown in this equation.

Pressure: Pressure is calculated using the relation

$$P = \frac{RT}{V-b} - \frac{a}{V(V+b)+b(V-b)}. \tag{2.18}$$

where V is the molar volume, a and b are the attraction and co-volume parameters, respectively. Both a and b are functions of temperature, and their formulations will be discussed shortly.

Compressibility factor: To calculate the compressibility factor, (2.18) is first rearranged as

$$\left(P + \frac{a}{V(V+b)+b(V-b)}\right)(V-b) = RT. \tag{2.19}$$

Now, comparing (2.17) and (2.19), we can write a cubic polynomial in Z as

$$Z^3 - (1-B)Z^2 + (A - 3B^2 - 2B)Z - (AB - B^2 - B^3) = 0. \tag{2.20}$$

In the above equation,

$$A = \frac{aP}{(RT)^2}, \tag{2.21a}$$

$$B = \frac{bP}{RT}. \tag{2.21b}$$

It must be noted that the number of roots of this polynomial is governed by the number of phases in the system.

If one applies (2.19) at critical temperature (T_c) and critical pressure (P_c), then we get

$$a|_{T=T_c} = 0.45724\frac{(RT_c)^2}{P_c}, \tag{2.22}$$

$$b|_{T=T_c} = 0.0778\frac{RT_c}{P_c}, \tag{2.23}$$

$$Z_c = 0.3070. \tag{2.24}$$

While $b = b|_{T=T_c}$ at other temperatures, a is calculated as

$$a(T) = a(T_c)\theta(T_r, \omega). \tag{2.25}$$

Thus, a is scaled via a dimensionless function of reduced temperature (T_r) and acentric factor (ω). The precise equation for θ is

$$\theta = \left[1 + \beta \left(1 - \sqrt{(T/T_c)} \right) \right]^2, \tag{2.26}$$

where, depending upon the range of ω

$$\beta = 0.37464 + 1.54226\omega - 0.26992\omega^2; 0 < \omega < 0.5, \tag{2.27}$$

$$\beta = 0.3796 + 1.485\omega - 0.1644\omega^2 + 0.01667\omega^3; 0.5 < \omega < 2.0. \tag{2.28}$$

It must be noted that this dimensionless function (θ) evaluates to 1 at $T = T_c$.

Fugacity coefficient: The fugacity of the component in the mixture is calculated using the relationship

$$\ln \frac{\phi_k}{x_k P} = \frac{b_k}{\sum_i x_i b_i} (Z - 1) - \ln(Z - B) - \frac{A}{2\sqrt{2}B} \tag{2.29}$$

$$\times \left(\frac{2\sum_j X_j a_{ik}}{\sum_i \sum_j x_i x_j a_{ij}} - \frac{b_k}{\sum_i x_i b_i} \right) \ln \left(\frac{Z + 2.414B}{Z - 0.414B} \right).$$

In the above equation, a_{ij} is related to an empirically determined binary interaction coefficient between the components i and j, viz., δ_{ij}, and is given as

$$a_{ij} = (1 - \delta_{ij})\sqrt{a_i a_j}. \tag{2.30}$$

Enthalpy: Enthalpy is calculated as

$$H = H_{\text{ideal}} + RT(Z - 1) + \frac{T\frac{da}{dT} - a}{2\sqrt{2}b} \ln \left(\frac{Z + 2.414B}{Z - 0.414B} \right), \tag{2.31}$$

where H_{ideal} is the enthalpy of the ideal gas at the same temperature and pressure conditions.

Volume translated Peng–Robinson equation of state: The *volume translated* Peng–Robinson (*v-PR*) equation of state is a slight modification of the *PR* equation of state to improve the volumetric predictions. In this, the following shift parameter, ε, is introduced [34]

$$\varepsilon = 1 - d/M^e. \tag{2.32}$$

In the above equation, d and e are constants that depend upon the mixture. For example, for n-alkanes, $d = 2.258$ and $e = 0.1823$. With this shift parameter, the corrected molar volume is

$$V_{\text{corrected}} = V_{\text{PR}} - \varepsilon x_i b_i. \tag{2.33}$$

As a result of this small modification, the v-PR equation of state predicts the equilibrium properties of the hydrocarbon mixtures quite accurately.

2.2.2.2 Perturbed Chain Statistical Associating Fluid Theory Equation of State

In Perturbed chain statistical associating fluid theory equation of state, proposed by Gross and Sadowski [22, 23], once again, the density is computed as

$$\rho = \frac{P}{ZRT} M. \tag{2.34}$$

In this expression, unlike PR equation of state, the compressibility factor is calculated using the relation

$$Z = 1 + \eta \left(\frac{\partial \tilde{a}^{\text{res}}}{\partial \eta} \right)_{T,c_i}, \tag{2.35}$$

where $\tilde{a}^{\text{res}}$ is the residual Helmoltz free energy and η is the packing fraction.

The residual Helmoltz free energy, $\tilde{a}^{\text{res}}$, that is used to compute the other necessary mixture properties, is the key factor in PC-$SAFT$ equation of state. It is represented as the sum of three contributing factors, viz., the hard chain ($\tilde{a}^{\text{hc}}$), dispersion ($\tilde{a}^{\text{disp}}$), and association ($\tilde{a}^{\text{assoc}}$). The hard chain contribution is computed as

$$\tilde{a}^{\text{hc}} = \bar{m} \tilde{a}^{\text{hs}} - \sum_i c_i (m_i - 1) \ln g_{ii}^{\text{hs}}(\sigma_{ii}), \tag{2.36}$$

where $\bar{m}$ is the mean segment number in the system, m_i is the number of segments per chain of the ith component, and σ_{ii} is the segment diameter. Additionally, in the above equation, the hard sphere fluid's Helmoltz free energy ($\tilde{a}^{\text{hs}}$), and the radial distribution function (g_{ij}^{hs}) are given as

$$\tilde{a}^{\text{hs}} = \frac{1}{\gamma_0} \left[\frac{3\gamma_1 \gamma_2}{(1 - \gamma_3)} + \frac{\gamma_2^3}{\gamma_3(1 - \gamma_3)^2} + \left(\frac{\gamma_2^3}{\gamma_3^2} - \gamma_0 \right) \ln(1 - \gamma_3) \right], \tag{2.37}$$

$$g_{ij}^{\text{hs}} = \frac{1}{(1 - \gamma_3)} + \left(\frac{d_i d_j}{d_i + d_j} \right) \frac{3\gamma_2}{(1 - \gamma_3)^2} + \left(\frac{d_i d_j}{d_i + d_j} \right)^2 \frac{2\gamma_2^2}{(1 - \gamma_3)^3}, \tag{2.38}$$

where γ_k represents

$$\gamma_k = \rho \pi / 6 \sum_i c_i m_i d_i^k. \tag{2.39}$$

In this relation, the temperature-dependent segment diameter, d_i, of the ith component is given in terms of the depth of the potential pair (ε_i) by

$$d_i = \sigma_i \left[1 - 0.12 \exp\left(-3\frac{\varepsilon_i}{kT} \right) \right]. \tag{2.40}$$

The dispersion contribution to the Helmoltz free energy is

$$\tilde{a}^{\text{disp}} = -2\pi\rho I_1(\eta,\bar{m})\overline{m^2\varepsilon\sigma^3} - \pi\rho\bar{m}C_1 I_2(\eta,\bar{m})\overline{m^2\varepsilon^2\sigma^3}, \tag{2.41}$$

where three abbreviations have been used, viz.

$$C_1 = \left(1 + \bar{m}\frac{8\eta - 2\eta^2}{(1-\eta)^4} + (1-\bar{m})\frac{20\eta - 27\eta^2 + 12\eta^3 - 2\eta^4}{(1-\eta)^2(2-\eta)^2} \right), \tag{2.42a}$$

$$\overline{m^2\varepsilon\sigma^3} = \sum_i \sum_j c_i c_j m_i m_j \left(\frac{\varepsilon_{ij}}{kT} \right) \sigma_{ij}^3, \tag{2.42b}$$

$$\overline{m^2\varepsilon^2\sigma^3} = \sum_i \sum_j c_i c_j m_i m_j \left(\frac{\varepsilon_{ij}}{kT} \right)^2 \sigma_{ij}^3. \tag{2.42c}$$

In the above equations, σ_{ij} and ε_{ij} are determined as

$$\sigma_{ij} = 0.5(\sigma_i + \sigma_j). \tag{2.43}$$

$$\varepsilon_{ij} = \sqrt{\varepsilon_i\varepsilon_j}(1 - k_{ij}), \tag{2.44}$$

k_{ij} being the binary interaction parameter. In (2.41), the integrals of the perturbation theory, I_1 and I_2, are approximated via the following expressions

$$I_1(\eta,\bar{m}) = \sum_{i=0}^{6} a_i(\bar{m})\eta^i, \tag{2.45}$$

$$I_2(\eta,\bar{m}) = \sum_{i=0}^{6} b_i(\bar{m})\eta^i. \tag{2.46}$$

The coefficients a_i and b_i in these expressions depend upon the chain length and are given as

$$a_i(\bar{m}) = a_{0,i} + a_{1,i}\frac{\bar{m}-1}{\bar{m}} + a_{2,i}\frac{\bar{m}-1}{\bar{m}}\frac{\bar{m}-2}{\bar{m}}, \tag{2.47}$$

$$b_i(\bar{m}) = b_{0,i} + b_{1,i}\frac{\bar{m}-1}{\bar{m}} + b_{2,i}\frac{\bar{m}-1}{\bar{m}}\frac{\bar{m}-2}{\bar{m}}, \tag{2.48}$$

where $a_{k,i}$ and $b_{k,i}$ are model constants.

Finally, the association contribution, $\tilde{a}^{\text{assoc}}$, is given by

$$\tilde{a}^{\text{assoc}} = \sum_i c_i \sum_{A_i} (\ln X_{A_i} - 0.5X_{A_i} + 0.5M_i), \tag{2.49}$$

where M_i is the number of associating sites on the molecule of type i and X_{A_i} is the fraction of A sites that do not form associating bonds with other active sites. Further, X_{A_i} is a solution of the following system of equations:

$$X_{A_i} = \frac{1}{1 + \rho \sum_j c_j \sum_{B_j} X_{B_j} \Delta^{A_i B_j}},$$ (2.50)

where B_j indicates the summation over all the sites and $\Delta^{A_i B_j}$ is a measure of the association strength between site A on molecule i and site B on molecule j. Now, if sites A and B are on the molecules of the same type, then

$$\Delta^{A_i B_j} = \sigma^3 g_H(d_i) \kappa^{A_i B_j} \left[\exp\left(\frac{\varepsilon^{A_i B_j}}{k_B T} \right) - 1 \right],$$ (2.51)

where $\kappa^{A_i B_j}$ is the association volume and $\varepsilon^{A_i B_j}$ is the association energy. On the other hand, if sites A and B are on molecules of different type

$$\Delta^{A_i B_j} = \sqrt{\Delta^{A_i B_i} \Delta^{A_j B_j}}.$$ (2.52)

Using the above relations, quantities such as density, pressure, fugacity, enthalpy, and entropy can be determined as derivatives of the residual Helmoltz energy. The computation of these quantities are done as follows:

Density: In order to compute the density at a given system pressure, an iterated approach is followed in which the reduced density (η) is adjusted to match the calculated pressure with the system pressure. Once a value for η is determined, the number density of molecules, ρ, is estimated as

$$\rho = 6/\pi\eta \left(\sum_i c_i m_i d_i \right)^{-1}.$$ (2.53)

The molar density $\hat{\rho}$ is then calculated as

$$\hat{\rho} = \frac{\rho}{6.022 \times 10^{23}} \left(10^{10} \frac{A^\circ}{m} \right) \left(10^{-3} \frac{kmol}{mol} \right).$$ (2.54)

Pressure: From the residual Helmoltz free energy, the compressibility factor can be determined as

$$Z = 1 + \eta \left(\frac{\partial \tilde{a}^{res}}{\partial \eta} \right)_{T, c_i}.$$ (2.55)

The pressure is then calculated in the units of Pa as

$$P = ZkT\rho \left(10^{10} \frac{A^\circ}{m} \right)^3.$$ (2.56)

Fugacity coefficient: The residual chemical potential, μ_k^{res}, is obtained from $\tilde{a}^{\text{res}}$ as

$$\frac{\mu_k^{\text{res}}(T,v)}{kT} = \tilde{a}^{\text{res}} + (Z-1) + \left(\frac{\partial \tilde{a}^{\text{res}}}{\partial c_k}\right)_{T,v,c_{j\neq k}} - \sum_{j=1}^{N}\left[c_j\left(\frac{\partial \tilde{a}^{\text{res}}}{\partial c_j}\right)_{T,v,c_{i\neq j}}\right]. \quad (2.57)$$

The fugacity coefficient, ϕ_k, is then obtained from the chemical potential as

$$\ln\phi_k(T,v) = \frac{\mu_k^{\text{res}}(T,v)}{kT} - \ln Z. \quad (2.58)$$

Enthalpy and entropy: From the derivative of the residual Helmoltz free energy with respect to temperature, the molar enthalpy, $\hat{h}^{\text{res}}$, is given by the relation

$$\hat{h}^{\text{res}} = RT\left[-T\left(\frac{\partial \tilde{a}^{\text{res}}}{\partial T}\right)_{\rho,c_i} + (Z-1)\right]. \quad (2.59)$$

The residual entropy at a given pressure and temperature condition is

$$\hat{s}^{\text{res}}(P,T) = -RT\left[\left(\frac{\partial \tilde{a}^{\text{res}}}{\partial T}\right)_{\rho,c_i} + \frac{\tilde{a}^{\text{res}}}{T}\right] + R\ln(Z). \quad (2.60)$$

Finally, from the above two equation, the Gibbs free energy, $\hat{g}^{\text{res}}(P,T)$, is defined as

$$\hat{g}^{\text{res}} = \hat{h}^{\text{res}} - T\hat{s}^{\text{res}}(P,T). \quad (2.61)$$

2.2.3   Binary Liquid Mixtures

A thermodiffusion theory for a multicomponent mixture has been outlined in Sect. 2.2. However, most of the experimental research is conducted on two or more lately on the three component systems. This is primarily because of the complexity of the inter-particle interactions that is not well understood even in such simple systems. Consequently, numerous models have been proposed over time for binary mixtures. Most of these models differ in the way the net heat of transport has been modeled. Additionally, some of the models also incorporate the effects of additional forces that contribute to the separation behavior in the non-isothermal mixtures.

   A detailed description of the formulation and derivation of every model in the literature is beyond the scope of this book. Instead, in the ensuing paragraphs, we present an overview of a subset of them, in chronological order, giving adequate references for the reader to pursue them in detail.

2.2.3.1 1955—Dougherty and Drickamer—Model 1

As per this model [10], based on the principles of nonequilibrium thermodynamics, the flux equation for the first component in a binary system, for instance, can be written as

$$J_1 = -\chi D_m \left[\nabla x_1 - \frac{\alpha_T x_1}{x_2} T \nabla T \right]. \tag{2.62}$$

At steady state, the fluxes J_1 and J_2 vanish. Combining this condition with (2.12) for a binary mixture, and noting from the Gibbs–Duhem relation at constant pressure and temperature that

$$\sum_i x_i d\mu_i = 0, \tag{2.63}$$

the thermodiffusion factor, α_T, for a binary system can be written as

$$\alpha_T = \frac{Q_2^* - Q_1^*}{x_1 \frac{\partial \mu_1}{\partial x_1}}. \tag{2.64}$$

Now, in their model, Dougherty and Drickamer [10] related the net heat of transport to the energy needed to create a hole (W_H) and the energy needed to occupy a hole (W_L), respectively. Their model also accounted for the different sizes and shapes of the molecules by considering the different fractions of the two molecules that create and occupy a hole. More precisely,

$$Q_1^* = W_{H1} - \psi_1 W_L \text{ and} \tag{2.65a}$$

$$Q_2^* = W_{H2} - \psi_2 W_L, \tag{2.65b}$$

where

$$\psi_i = \frac{V_i}{x_1 V_1 + x_2 V_2}, \tag{2.66a}$$

$$W_L = x_1 W_{H_1} + x_2 W_{H_2} \text{ and} \tag{2.66b}$$

$$W_{Hi} = -\frac{1}{\tau_i} \left(U_i - U_{ig} \right). \tag{2.66c}$$

In the above equation, τ is the ratio of the energy of vaporization of the liquid and the activation energy of the viscous flow, U_i is the partial molar energy of the ith component, and U_{ig} is the energy of the ideal gas at the corresponding temperature and pressure.

2.2.3.2 1955—Dougherty and Drickamer—Model 2

In this model, the authors proposed a slightly different expression for α_T as [11]

$$\alpha_T = \frac{M_1 Q_2^* - M_2 Q_1^*}{(M_1 x_1 + M_2 x_2) x_1 \frac{\partial \mu_1}{\partial x_1}}. \tag{2.67}$$

Unlike the previous model where the center of mass of the system is assumed stationary, the above expression avoids this hidden assumption. Now, expressing the net heats of transport in terms of W_{Hi} and W_L, α_T is calculated as

$$\alpha_T = \frac{M_2 V_1 + M_1 V_2}{(M_1 x_1 + M_2 x_2) x_1 \frac{\partial \mu_1}{\partial x_1}} \left(\frac{W_{H2}}{V_2} - \frac{W_{H1}}{V_1} \right). \tag{2.68}$$

It must be noted that the convention in the Drickamer models is that if $\alpha_T > 0$ for component-1, then it enriches near the hot side.

2.2.3.3 1956—Tichacek, Kmak, and Drickamer

By discussing the thermodiffusion process in a center-of-volume frame of reference, the authors proposed an expression for α_T as [54]

$$\alpha_T = \frac{V_1 V_2}{(V_1 x_1 + V_2 x_2) x_1 \frac{\partial \mu_1}{\partial x_1}} \left(\frac{M_2 Q_2^*}{V_2} - \frac{M_1 Q_1^*}{V_1} \right). \tag{2.69}$$

2.2.3.4 1969—Haase

While Haase [26] studied the barodiffusion process in isothermal gas mixtures subjected to pressure gradients, he proposed an expression for thermodiffusion in binary electrolyte mixtures, by analogy, as

$$\alpha_T = \frac{M_1 (H_2 - H_2^{(0)}) - M_2 (H_1 - H_1^{(0)})}{(M_1 x_1 + M_2 x_2) x_2 \frac{\partial \mu_2}{\partial x_2}} + \frac{RT}{x_2 \frac{\partial \mu_2}{\partial x_2}} \alpha_0, \tag{2.70}$$

where H is the partial molar enthalpy and α_0 is the thermodiffusion factor for the ideal gas mixture at the corresponding temperature. In the above equation, the first term on the right side is the thermodiffusion factor of the mixture with respect to the ideal gas phase that is indicated by the superscript (0), and the second term is the contribution due to the thermodiffusion in an ideal gas state.

2.2.3.5 1986—Guy

In a barycentric frame of reference, Guy [25] proposed the following expression for thermodiffusion factor in electrolyte solutions:

$$\alpha_T = \frac{M_2\left(h_2^{xs} - h_1^{xs}\right)}{x_2 \frac{\partial \mu_2}{\partial x_2}}, \tag{2.71}$$

where h_i^{xs} is the partial excess enthalpy of the ith component in the mixture.

2.2.3.6 1989—Kempers

Following an initial model [35] based on the principles of statistical nonequilibrium thermodynamics, that is similar to the Haase model, in 2001, Kempers [36] proposed revised expressions for thermodiffusion factor in multicomponent systems. In particular, two expressions of his work that correspond to the *center-of-volume* and *center-of-mass* frames of reference, applicable to binary systems, are as follows:

Center of volume frame of reference

$$\alpha_T = \left(\frac{V_1 V_2}{V_1 x_1 + V_2 x_2}\right) \frac{\frac{(H_2 - H_2^{(0)})}{V_2} - \frac{(H_1 - H_1^{(0)})}{V_1}}{x_1 \frac{\partial \mu_1}{\partial x_1}} + \frac{RT}{x_1 \frac{\partial \mu_1}{\partial x_1}} \alpha_{T0}. \tag{2.72}$$

Center of mass frame of reference

$$\alpha_T = \left(\frac{M_1 M_2}{M_1 x_1 + M_2 x_2}\right) \frac{\frac{(H_2 - H_2^{(0)})}{M_2} - \frac{(H_1 - H_1^{(0)})}{M_1}}{x_1 \frac{\partial \mu_1}{\partial x_1}} + \frac{RT}{x_1 \frac{\partial \mu_1}{\partial x_1}} \alpha_{T0}. \tag{2.73}$$

In the above equations, α_{T0} is the thermodiffusion factor of the mixture at the corresponding temperature in the ideal gas state.

2.2.3.7 1998—Shukla and Firoozabadi

Coupling the model of Drickamer with the volume translated Peng–Robinson equation of state, Shukla and Firoozabadi proposed a model for α_T in binary liquid mixtures as [50]

$$\alpha_T = \frac{\frac{U_1}{\tau_1} - \frac{U_2}{\tau_2}}{x_1 \frac{\partial \mu_1}{\partial x_1}} + \frac{(V_2 - V_1)\left(x_1 \frac{U_1}{\tau_1} + x_2 \frac{U_2}{\tau_2}\right)}{(x_1 V_1 + x_2 V_2) x_1 \frac{\partial \mu_1}{\partial x_1}}. \tag{2.74}$$

It is worth noting that the sign convention in this expression is that when $\alpha_T > 0$ for component-1, it enriches near the cold side.

2.2.3.8   2008—Artola, Rousseau, and Galliero

Following the propositions of Prigogine [44], Artola et al. [1] proposed the following expression based on the principles of nonequilibrium thermodynamics:

$$\alpha_T = \frac{\Delta G_2 - \Delta G_1}{RT} + \frac{M_2 - M_1}{M_1 + M_2} \frac{\Delta G_1 + \Delta G_2}{RT}. \tag{2.75}$$

In the above equation, ΔG_i is the activation-free enthalpy of the ith component and is obtained as

$$D_i = D_{i0} \exp(-\Delta G_i / RT). \tag{2.76}$$

In this equation, D_i and D_{i0} are the self diffusion coefficients at current temperature T and reference temperature T_0, respectively.

2.2.3.9   Other Expressions Based on Activation Energy of Viscous Flow

Several other models have evolved in the study of thermodiffusion, which are particularly influenced by the work of Tichacek et al. [54], modeling the net heat of transport via an activation energy of viscous flow (E^{vis}). While a universal model for thermodiffusion is still lacking in the literature, as outlined below, one can find different expressions for α_T for different types of mixtures, viz., nonassociating liquids [12], associating liquids [13], polymers [17], and DNA solutions [18].

Nonassociating liquid mixtures, e.g., hydrocarbon liquid mixtures [12]:

$$\alpha_T = \frac{E_1^{vis} - E_2^{vis}}{x_1 \frac{\partial \mu_1}{\partial x_1}}, \tag{2.77}$$

$$\alpha_T = \frac{M_2 E_1^{vis} - M_1 E_2^{vis}}{(M_1 x_1 + M_2 x_2) x_1 \frac{\partial \mu_1}{\partial x_1}}, \tag{2.78}$$

$$\alpha_T = \frac{V_2 E_1^{vis} - V_1 E_2^{vis}}{(V_1 x_1 + V_2 x_2) x_1 \frac{\partial \mu_1}{\partial x_1}}. \tag{2.79}$$

In addition to the above expressions, one can also use a weighted combination of (2.78) and (2.79) as a measure of α_T.

Associating liquid mixtures, e.g., water–alcohol mixtures [13]:

$$\alpha_T = \frac{M_2 \left(E_1^{vis} - E_{mix}^{vis} \right) - M_1 \left(E_2^{vis} - E_{mix}^{vis} \right)}{(M_1 x_1 + M_2 x_2) x_1 \frac{\partial \mu_1}{\partial x_1}}, \tag{2.80}$$

$$\alpha_T = \frac{V_2 \left(E_1^{\text{vis}} - E_{\text{mix}}^{\text{vis}} \right) - V_1 \left(E_2^{\text{vis}} - E_{\text{mix}}^{\text{vis}} \right)}{(V_1 x_1 + V_2 x_2) x_1 \frac{\partial \mu_1}{\partial x_1}}. \tag{2.81}$$

While these two equations are not particularly good in their predictions, an equation that performs somewhat better is given as

$$\alpha_T = -\frac{E_1^{\text{vis}} - E_2^{\text{vis}}}{x_1 \frac{\partial \mu_1}{\partial x_1}} \frac{\partial \ln \left(\eta_{\text{mix}} / \eta_0 \right)}{\partial x_1}, \tag{2.82}$$

where η_0 is a reference viscosity, e.g., viscosity of water (1 cP at room temperature). In this last expression, by including a term containing the derivative of the natural logarithm of viscosity with respect to the composition, there is an attempt to account for the nonlinear relation between the mixture viscosity and composition, which is often observed in associating mixtures. Thus, by including this term, the sign change in the thermodiffusion factor has been forced at the maxima or minima on the $\ln(\eta)$ versus x_1 graph.

Dilute polymer and DNA solutions [18]:

$$\alpha_T = \frac{E_M^{\text{vis}} a \ln \left(\frac{2M_P}{M_M} \right) - E_S^{\text{vis}}}{RT}, \tag{2.83}$$

where M_P and M_M are the molecular weights of the polymer and monomer, respectively. a is a polymer-solvent characteristic parameter that varies between 0.5 and 0.8. E_S^{vis} and E_M^{vis} are the activation energy of viscous flow of the solvent and the monomer, respectively.

2.2.3.10 A Note on the Activation Energy of Viscous Flow

As mentioned at the beginning of Sect. 2.2.3.9, in (2.77)–(2.83) the net heat of transport Q_i^* for the ith component has been calculated by correlating it with activation energy of viscous flow. In this, by applying Eyring's rate theory [20], viscosity of a liquid can be related to the activation energy of the viscous flow. In particular, in the vicinity of the temperature of interest, if we plot the natural logarithm of the experimental data of viscosity or its product with the molar volume against $1/RT$, the graph exhibits a straight line with a slope corresponding to the activation energy of viscous flow. More precisely

$$E_i^{\text{vis}} \approx \frac{\partial \ln \eta}{\partial (1/T)}. \tag{2.84}$$

This calculation of activation energy of viscous flow is applicable for nonassociating mixtures like liquid hydrocarbons.

Fig. 2.1 Natural logarithm of viscosity plotted against the mole fraction of the first component. The trends are for the experimental data obtained from [51] for associating (water–alcohol) mixtures and from [3] for nonassociating (hydrocarbon) mixtures. IBB is isobutylbenzene, nC_{12} is n-dodecane, and THN is tetralin

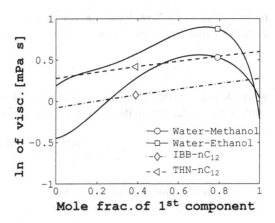

For the activation energy of viscous flow for the mixture (E_{mix}^{vis}) in (2.80)–(2.82), the above correlation cannot be used. This is because in associating mixtures like water–alcohol, for instance, there is a nonlinear relationship between $\ln \eta$ and $1/RT$. This is clearly evident in Fig. 2.1 where the natural logarithm of viscosity for two representative water–alcohol mixtures are plotted alongside the viscosity data of two hydrocarbon mixtures. The reason behind the nonlinear trend in these associating mixtures is that with the change in the concentration of the liquid, the interactions between the molecules and the structure of the mixtures change. One way to account for these changes is to simply multiply the expressions of α_T by a correction term, E_{mix}^{vis}, which is prescribed as [13]

$$E_{mix}^{vis} \approx \frac{\partial \ln (\eta_{mix}/\eta_0)}{\partial x}. \tag{2.85}$$

This gives us the corrected α_T as in (2.82). Note that the above equation is also incorporated in (2.80) and (2.81).

The calculation of the activation energy of viscous flow in the polymer solutions is even more complicated. This is because polymers are in solid state at room temperature although they are soluble in certain liquids, resulting in a liquid mixture. Now, with most of the thermodiffusion experiments on polymer solutions being performed at room temperatures, viscosity data of pure polymer at room temperature is unavailable.

On the other hand, what is known of polymers is that in dilute polymer solutions, small sections of these long chain molecules called *beads* or *segments* move independent of each other. It must be noted that this is valid only for a very low concentration of the polymer in the mixture. So, in this case, we can use the activation energy of the viscous flow of the monomers that represent the moving segments, instead of the activation energy of the polymer. This is something that can be calculated since the monomers are usually in liquid state even at room temperatures. Now, for the activation energy, viscosity data are needed. One correlation that can be used is the Mark–Houwink equation [30],

$$\eta = KM_p^a, \tag{2.86}$$

where K is a constant dependent on the polymer-solvent system. Using the above expression, we can obtain the activation energy of viscous flow of the polymer as

$$E_P^{vis} = \frac{d\ln(V\eta)}{d(1/RT)}\ln\left(\frac{M_P}{M_B}\right)^a, \tag{2.87}$$

where M_B is the molecular weight of the moving segment of the polymer and the other notations have the usual meaning, as defined before.

2.2.3.11   Other Theoretical Formulations

While the nonequilibrium thermodynamic approaches using the net heats of transport have been a fairly popular approach to study thermodiffusion and formulate theoretical equations for this phenomenon in different mixtures, there are other theories for the thermodiffusive separation.

A hydrodynamic approach: In this approach to study thermodiffusion, one can express the thermodiffusional driving force as a combination of equilibrium and nonequilibrium terms [41]. The equilibrium term can be attributed to the temperature gradient of the partial pressure. Further, by applying the solvation theory to the inter-particle interactions, one can write

$$S_T = \frac{1}{n_1\kappa_B T}\left(\frac{\partial p_1}{\partial T}\right)_{p,c_1\to 0}, \tag{2.88}$$

where n_1, p_1, and c_1 are the number of moles, partial pressure, and the concentration of the first component, i.e., the solute.

By theoretically defining the partial pressure, Morozov [41] deduced two useful expressions for the Soret coefficient as

$$S_T = \frac{1}{n_2 T}\left[\frac{\partial\,(n_2 T Z_{12})}{\partial T}\right]_p, \tag{2.89}$$

$$S_T = \frac{1}{n_2 T}\left[\frac{\partial\,(n_2 T\,[Z_{12}-Z_2])}{\partial T}\right]_p. \tag{2.90}$$

In the above equations, n_2 is the number of moles of solvent, Z_{12} is the compressibility factor of the first component in the solvent, and Z_2 is the compressibility factor of the solvent.

Statistical thermodynamics approach: By considering thermodiffusion on the basis of statistical thermodynamics and force balance, one can express the thermodiffusion coefficient in terms of the pair-interaction potential between the colloidal spheres [9]. More precisely,

$$D_T = D_0 \beta \frac{\partial \Pi}{\partial T}, \tag{2.91}$$

where D_0 is the Einstein's translational diffusion coefficient, $\beta = 1/\kappa_B T$, κ_B being the Boltzmann constant, and Π is the osmotic pressure. This osmotic pressure is a function of the colloid number density, temperature, and the chemical potential of the pure solvent with which the suspension is in osmotic equilibrium, and is given as

$$\Pi = \rho \kappa_B T - \frac{2\pi}{3} \rho^2 \int_0^\infty r^3 g^{eq} \frac{df}{dr} dr, \tag{2.92}$$

where r is the distance between the two colloidal spheres and f is the pair-interaction potential of mean force. g^{eq} is the equilibrium pair-correlation function which for low colloidal concentrations is given as

$$g^{eq} = \exp(-\beta f). \tag{2.93}$$

2.2.4 Ternary Liquid Mixtures

Recent advances in experimental techniques has resulted in experimental investigations on thermodiffusion in ternary liquid mixtures. Theoretical framework for studying ternary mixtures has been around for much longer though. Of course, the general multicomponent formulation of Sect. 2.2 with $n = 3$ will give an expression for the ternary mixture as well. Nevertheless, there are some formulations for the thermodiffusion coefficients in ternary mixtures in the literature. Here we present two sets of expressions for the thermodiffusion coefficients in ternary mixtures, one for nonassociating liquids and the other for associating liquids, as outlined by Eslamian and Saghir [14].

The starting point of the derivation for both associating and nonassociating liquid mixtures is the molar flux of the ith component in a n-component mixture that can be written as

$$\mathbf{J}_i = -\sum_{k=1}^n L_{ik} \left[Q_k^* \frac{\nabla T}{T} + \sum_{j=1}^{n-1} \frac{\partial \mu_k}{\partial x_j} \nabla x_j \right], \quad i = 1, \ldots, n. \tag{2.94}$$

At steady state, for a ternary mixture ($n = 3$), the flux of each component is zero, i.e., $\mathbf{J}_1 = \mathbf{J}_2 = \mathbf{J}_3 = 0$. In other words, we have

$$\frac{\partial \mu_1}{\partial x_1} \nabla x_1 + \frac{\partial \mu_1}{\partial x_2} \nabla x_2 = -Q_1^* \frac{\nabla T}{T}, \tag{2.95a}$$

$$\frac{\partial \mu_2}{\partial x_1} \nabla x_1 + \frac{\partial \mu_2}{\partial x_2} \nabla x_2 = -Q_2^* \frac{\nabla T}{T}, \tag{2.95b}$$

$$\frac{\partial \mu_3}{\partial x_1} \nabla x_1 + \frac{\partial \mu_3}{\partial x_2} \nabla x_2 = -Q_3^* \frac{\nabla T}{T}. \tag{2.95c}$$

In writing the above system of equations, it is noted that the phenomenological coefficients are nonzero and can therefore be eliminated. Further, since the fluxes of only two components are independent, we can eliminate the third equation in the above system using the Gibbs–Duhem relation at constant temperature and pressure conditions, and writing $\frac{\partial \mu_3}{\partial x_1}$ and $\frac{\partial \mu_3}{\partial x_2}$ in terms of the chemical potentials and mole fractions of the first two components. This modifies the above system of equations as

$$B_{11} \nabla x_1 + B_{12} \nabla x_2 = -(Q_1^* - Q_3^*) \frac{\nabla T}{T}, \tag{2.96a}$$

$$B_{21} \nabla x_1 + B_{22} \nabla x_2 = -(Q_2^* - Q_3^*) \frac{\nabla T}{T}, \tag{2.96b}$$

where B_{ij} are elements of the matrix $\mathbf{B}$, and are given as

$$B_{11} = \left[\frac{\partial \mu_1}{\partial x_1} \left(1 + \frac{x_1}{x_3} \right) + \frac{\partial \mu_2}{\partial x_1} \left(\frac{x_2}{x_3} \right) \right], \tag{2.97a}$$

$$B_{12} = \left[\frac{\partial \mu_1}{\partial x_2} \left(1 + \frac{x_1}{x_3} \right) + \frac{\partial \mu_2}{\partial x_2} \left(\frac{x_2}{x_3} \right) \right], \tag{2.97b}$$

$$B_{21} = \left[\frac{\partial \mu_2}{\partial x_1} \left(1 + \frac{x_2}{x_3} \right) + \frac{\partial \mu_1}{\partial x_1} \left(\frac{x_1}{x_3} \right) \right], \tag{2.97c}$$

$$B_{22} = \left[\frac{\partial \mu_2}{\partial x_2} \left(1 + \frac{x_2}{x_3} \right) + \frac{\partial \mu_1}{\partial x_2} \left(\frac{x_1}{x_3} \right) \right]. \tag{2.97d}$$

To calculate the derivatives of the chemical potential one can employ an appropriate equation of state to first calculate the fugacity and its derivative for the ith component. Subsequently, these values can be used to calculate $\frac{\partial \mu_i}{\partial x_j}$ using the relation

$$\frac{\partial \mu_i}{\partial x_j} = \frac{RT}{f_i} \frac{\partial f_i}{\partial x_j}. \tag{2.98}$$

As noted previously, one can write the flux equations using the transport coefficients as

$$\mathbf{J}_1 = -\chi \left(D_{11} \nabla x_1 + D_{12} \nabla x_2 + \mathscr{D}_T^1 \nabla T \right), \tag{2.99a}$$

$$\mathbf{J}_2 = -\chi \left(D_{21} \nabla x_1 + D_{22} \nabla x_2 + \mathscr{D}_T^2 \nabla T \right), \tag{2.99b}$$

where D_{ij} are the components of the 2×2 matrix for molecular diffusion, viz., **D**. Also, it must be noted that for the above system of equations, $\sum_i \mathscr{D}_T^i = 0$. Once again, at steady state, in the absence of any flux, the left side of the above system can be set to zero and we can compare the above equations to (2.96) to obtain the thermodiffusion coefficients as

$$\mathscr{D}_T^1 = \frac{d(Q_1^* - Q_3^*) - b(Q_2^* - Q_3^*)}{T(ad - bc)}, \tag{2.100a}$$

$$\mathscr{D}_T^2 = \frac{-c(Q_1^* - Q_3^*) + a(Q_2^* - Q_3^*)}{T(ad - bc)}, \tag{2.100b}$$

where

$$a = \frac{B_{11}D_{22} - B_{12}D_{21}}{|\mathbf{D}|}, \tag{2.101a}$$

$$b = \frac{-B_{11}D_{12} + B_{12}D_{11}}{|\mathbf{D}|}, \tag{2.101b}$$

$$c = \frac{B_{21}D_{22} - B_{22}D_{21}}{|\mathbf{D}|}, \tag{2.101c}$$

$$d = \frac{-B_{21}D_{12} + B_{22}D_{11}}{|\mathbf{D}|}. \tag{2.101d}$$

Nonassociating mixtures: For the nonassociating mixtures like liquid hydrocarbons, in (2.100) we can use the activation energy of viscous flow as a measure of the net heat of transport, i.e.,

$$Q_i^* = E_i^{\text{vis}}. \tag{2.102}$$

This will get us the thermodiffusion coefficients for the ternary mixture as[1]

$$\mathscr{D}_T^1 = \frac{d(E_1^{\text{vis}} - E_3^{\text{vis}}) - b(E_2^{\text{vis}} - E_3^{\text{vis}})}{T(ad - bc)}, \tag{2.103a}$$

$$\mathscr{D}_T^2 = \frac{-c(E_1^{\text{vis}} - E_3^{\text{vis}}) + a(E_2^{\text{vis}} - E_3^{\text{vis}})}{T(ad - bc)}. \tag{2.103b}$$

Associating mixtures: As in the case of the expressions for the binary associating mixtures, the contributions due to the nonlinear relationship between the natural logarithm of viscosity and the composition of the mixture must be taken into account even in the ternary formulations. This can be done by multiplying each $(E_i^{\text{vis}} - E_j^{\text{vis}})$

---

[1] We would like to caution the readers that in [15] these equations were validated using an incorrect experimental data. The authors inadvertently used a wrong sign of the experimental data in the validation of these expressions with respect to the ternary hydrocarbon mixtures of n-dodecane-isobutylbenzene-tetralin.

term in (2.103) by a factor corresponding to the rate of change of the viscosity of a binary mixture of components i and j with respect to the mole fraction of component i. As a result, (2.103) gets modified for a ternary mixture of associating liquids as [16]

$$\mathscr{D}_T^1 = \frac{d(E_1^{vis} - E_3^{vis})\left(\frac{-\partial \ln(\eta_{13}/\eta_0)}{\partial x_1}\right) - b(E_2^{vis} - E_3^{vis})\left(\frac{-\partial \ln(\eta_{23}/\eta_0)}{\partial x_2}\right)}{T(ad - bc)}, \quad (2.104a)$$

$$\mathscr{D}_T^2 = \frac{-c(E_1^{vis} - E_3^{vis})\left(\frac{-\partial \ln(\eta_{13}/\eta_0)}{\partial x_1}\right) + a(E_2^{vis} - E_3^{vis})\left(\frac{-\partial \ln(\eta_{23}/\eta_0)}{\partial x_2}\right)}{T(ad - bc)}. \quad (2.104b)$$

In the above equation, the coefficients a, b, c, and d are as described in (2.101). η_{ij} is the viscosity of the binary mixture of component i and j.

2.3 Method 2: Simple Algebraic Expressions to Quantify Thermodiffusion

In the previous section, a detailed formalism of the thermodynamic approaches to study thermodiffusion in multicomponent mixtures has been presented. The numerous expressions presented, especially the ones requiring extensive thermo-dynamic data, need to be coupled to an appropriate equation of state to obtain estimates of quantities such as enthalpy, fugacity (to calculate chemical potential), and compressibility factor (for density calculation). On the other hand, following years of research on specific types of mixtures, simpler algebraic equations have been developed which can give a fairly accurate quantitative estimate of the thermodiffusion process. These are discussed in the ensuing paragraphs.

2.3.1 Liquid Hydrocarbon Mixtures

Several expressions are found in the analysis of the experimental data of different nonassociating mixtures, hydrocarbons in particular. One type of analysis procedure is to choose a binary mixture of two components and study the thermodiffusion in this mixture at different composition, or different temperatures. Such analyses cast light on the composition and temperature effects in the thermodiffusion process. Another type of analysis is to keep one component fixed and change the second component. This study, for instance, can yield information on the effect of the relative molecular weights of the participating components.

Several empirical correlations have been proposed after studying numerous binary systems. The correlations presented are for the Soret coefficient or for the thermodiffusion coefficient. A very often postulated relation for the Soret coefficient is

$$S_T \propto T^{-2}. \tag{2.105}$$

However, this power law is not true for all mixtures.

One correlation for a binary hydrocarbon mixture suggests that the Soret coefficient can be expressed in terms of composition- and temperature-dependent functions as [56]

$$S_T(x,T) = \theta(x)\phi(T) + S_T^i, \tag{2.106}$$

where $\theta(x)$ is a composition-dependent function, $\phi(T)$ is a temperature-dependent amplitude factor, and S_T^i is a constant offset that is independent of composition as well as temperature. These composition- and temperature-dependent functions are polynomials that are written as

$$\theta(x) = a_0 + a_1 x^2 + a_3 x^3 + \cdots, \tag{2.107a}$$
$$\phi(T) = 1 + b_1(T - T_0) + b_2(T - T_0)^2 + \cdots. \tag{2.107b}$$

In the above equations, a_i and b_i are constants that are empirically determined and T_0 is a reference temperature. Further, the degree of these polynomials would depend on the type of binary mixture studied, and is also determined empirically based on some experimental data about the mixture of interest.

Another such closely related algebraic formulation is [29]

$$S_T(x,T) = \theta(x)\phi(T) + c\frac{M_2 - M_1}{M_2 M_1}, \tag{2.108}$$

where c is a constant with a unit of K^{-1}. In addition to the composition and the operating temperature, this relation also incorporates the difference between the two participating molecules via the molecular weight term.

In a binary hydrocarbon mixture of two very similar molecules, if one investigates the isotopic effects by introducing varying degrees of deuteration, for instance, we can present the Soret coefficient as a sum of two contributions. The first contribution is due to the difference between the mass and the moment of inertia of the two components. This contribution is independent of the composition of the mixture and is attributed to the isotopic effect. The second term arises from the *chemical* contribution and is a function of composition. Thus, the Soret coefficient can be represented as [7]

$$S_T = S_T^c + (a\delta M + b\delta I), \tag{2.109}$$

where a (K^{-1}) and b (K^{-1}) are empirically determined coefficients. S_T^c (K^{-1}) is also empirically determined by regression analysis of the experimental data and has a form

$$S_T^c = (m_1 x + m_2) m_0, \tag{2.110}$$

where m_0 (K^{-1}), m_1 and m_2 are constants, and x is the mole fraction of the first component. δM and δI correspond to the effects of mass and moment of inertia, respectively, and are given as

$$\delta M = \frac{M_2 - M_1}{M_1 + M_2}, \tag{2.111a}$$

$$\delta I = \frac{I_2 - I_1}{I_1 + I_2}, \tag{2.111b}$$

where I_i is the moment of inertia of the ith component.

It must be noted that (2.109) only valid over a very narrow range of molecular masses and moments of inertia. A slight modification to this equation, making it accurate for a much wider range, has been proposed by Wittko and Köhler [55] as

$$S_T = S_T^c + (a\Delta M + b\Delta I), \tag{2.112}$$

where

$$\Delta M = M_2 - M_1, \tag{2.113a}$$

$$\Delta I = I_2 - I_1. \tag{2.113b}$$

By expressing the Soret coefficient as a function of the *absolute* difference instead of the *relative* difference of molecular weights and the moments of inertia, the range of equation has been extended.

In equimolar binary n-alkane mixtures, a quantitative estimate of the thermodiffusion coefficient can be obtained via a simple algebraic formulation that is essentially a polynomial in relative molecular weight. This is based on the fact that a simple quadratic relation exists between the thermodiffusion coefficient and the relative molecular weight, as shown in Fig. 2.2. This correlation is expressed as [2]

$$D_T = D_{T0}\delta M (1 + \lambda \delta M), \tag{2.114}$$

where D_{T0} is a reference thermodiffusion coefficient that is different for different n-alkane series and λ is a mixture-specific constant.

Algebraic correlations for the thermodiffusion coefficients can also be written in terms of the mixture thermodynamic properties such as dynamic viscosity and thermal expansion coefficients. For the equimolar binary n-alkane series considered in Fig. 2.2, viz., nC$_{10}$−nC$_i$, nC$_{12}$−nC$_i$, and nC$_{18}$−nC$_i$, Blanco et al. [2] have noted that there is a linear relation between the absolute molecular weight difference and Ω, i.e., $\Delta M \propto \Omega$, where Ω is defined as

$$\Omega = \frac{c_2 c_r D_T \eta}{\beta}, \tag{2.115}$$

Fig. 2.2 A quadratic trend between the thermodiffusion coefficients reported in [2] and the relative molecular weight in (2.111a) for equimolar binary n-alkane mixtures. nC_{10} is n-decane, nC_{12} is n-dodecane, and nC_{18} is n-octadecane. Figure modified from [2]

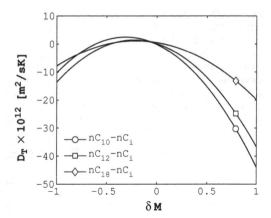

where the subscripts r and 2 are for the reference species and the second species, respectively, and β is the thermal expansion coefficient. A linear relationship between ΔM and Ω implies that we can write the thermodiffusion coefficient for the n-alkane mixtures as

$$D_T = K(M_r - M_2)\frac{\beta}{\eta c_r c_2}, \tag{2.116}$$

where K (m s^{-2}) is the proportionality constant.

To understand the formulation of (2.116), two points are worth noting:

1. The extent of the separation due to thermodiffusion in a binary mixture is governed by the amount of *similarity* between the molecules. The larger the similarity, the smaller the separation process and vice versa. A measure of the similarity between the molecules is the difference between their molecular weights (defined by ΔM in (2.113a)). This concept of similarity between the molecules is shown in Fig. 2.3 where $\Delta M = M_2 - M_r$ is plotted against D_T. We can see that as the difference between the molecular weights increases, the thermodiffusion coefficient increases.

2. A closer look at the figure reveals that D_T increases rapidly as the molecular weight of the second species (M_2) becomes much smaller than the reference species ($\Delta M \ll 0$). However, if the molecular weight of the second species is much larger than the molecular weight of the reference species ($\Delta M \gg 0$), the value of D_T is not very large. This is despite a large dissimilarity between the molecules. This is because when $\Delta M \gg 0$, the viscosity of the mixture increases, leading to decreased mobility of the species. At this point, if we recall the activation energy of viscous flow concept used to quantify thermodiffusion, it is noted that the energy needed to set the molecules in motion is very high for highly viscous fluids. So, in a highly simplistic algebraic model, one can use the viscosity of the mixture account for the mobility of the species in the mixture.

Fig. 2.3 The relation between the difference of the molecular weights of the two components in the binary mixture and the thermodiffusion coefficient of several binary n-alkane mixtures that are reported in [2]

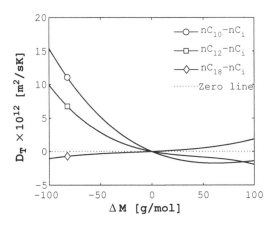

In (2.116), the similarity between the molecules is reflected in the term $(M_r - M_2)$ and the mobility is represented quantitatively by the inverse of viscosity. Of course, for equimolar mixtures,

$$c_r c_2 = \frac{M_r M_2}{(M_r + M_2)^2},\tag{2.117}$$

which transforms (2.116) to

$$D_{\mathrm{T}} = K(M_r - M_2)(M_r + M_2)^2 \frac{\beta}{\eta M_r M_2}.\tag{2.118}$$

Two issues with (2.114) and (2.116) are:

1. The constants in these equations are mixture specific. Hence, before employing these expressions, prior experimental data are needed to calculate these constants.
2. These equations are only valid for equimolar mixtures.

The first problem can be fixed by extending (2.114) for non-equimolar mixtures by introducing a correction term. More precisely, by including a linear or quadratic term to account for the effect of the mole fraction of the species, we can write two equations for the thermodiffusion coefficient in binary n-alkane mixtures as

$$D_{\mathrm{T}} = D_{\mathrm{T}0}\delta M \left(1 + \lambda_1 \delta M + \lambda_2 (x_r - x_2)\right),\tag{2.119a}$$

$$D_{\mathrm{T}} = D_{\mathrm{T}0}\delta M \left(1 + \lambda_1 \delta M + \lambda_2 (x_r - x_2) + \lambda_3 (x_r - x_2)^2\right),\tag{2.119b}$$

where x_r and x_2 are the mole fraction of the reference species and the second species, respectively. λ_i are series specific dimensionless constants. As before, $D_{\mathrm{T}0}$ is a reference thermodiffusion coefficient that is different for different binary n-alkane series. Note that for an equimolar case, these expressions reduce to (2.114).

Despite the aforementioned formulation in (2.119), making it valid for a much wider composition, we still need experimental data to determine the various

Table 2.1 Values of D_{T0} and λ_i in (2.121) and (2.122)

Equation #	$D_{T0} \times 10^{17}$ $[m^2 s^{-1} K^{-1}]$	λ_1	λ_2	$\lambda_3 \times 10^4$ $[mol\ g^{-1}]$
(2.121)	4.905	−0.114	−0.122	−3.27
(2.122)	1.213	−0.094	−1.053	−1.86

These constants are the same for any binary n-alkane mixture of any composition

constants in this equation. This drawback is unlikely to be fixed in this enhanced formulation that presents thermodiffusion coefficients solely in terms of the molecular weights and composition. On the other hand, by pursuing (2.116) that takes into account the mixture properties, one can extend/modify it to be valid for non-equimolar mixtures and also avoid the need for experimental data to determine any constants. To this end, an empirical equation proposed by Madariaga et al. [39] is

$$D_T = K_0(x_r)(M_r - M_2)\frac{\beta}{\eta c_r c_2}, \qquad (2.120)$$

where $K_0(x_r) = (5.34x_r - 7x_r^2 + 1.65x_r^3) \times 10^{-14}\ m\,s^{-2}$.[2]

Another formulation that is also valid for a wide range of n-alkane series of any composition and which does not need any prior data to determine the model constants is

$$D_T = D_{T0}\left(x_r + \lambda_1\Delta x + \lambda_2(\Delta x)^2 + \lambda_3\Delta M^{mix}\right)\frac{\beta\Delta M^{mix}}{\eta c_r c_2}, \qquad (2.121)$$

where $\Delta x = x_r - x_2$ and $\Delta M^{mix} = M_r - M_{mix}$, M_{mix} being the molecular weight of the mixture. A slight variation of the above equation is

$$D_T = D_{T0}\left(x_r + \lambda_1\Delta x + \lambda_2(\Delta x)^2 + \lambda_3\Delta M\right)\frac{\beta\Delta M}{\eta c_r c_2}. \qquad (2.122)$$

All three formulations ((2.120), (2.121), and (2.122)) are very accurate in predicting the thermodiffusion coefficients in any binary n-alkane mixture. The constants D_{T0} and λ_i in the latter two expressions are summarized in Table 2.1. Again, it must be emphasized that these constants in (2.121) and (2.122) do not change with the n-alkane series.

---

[2]In the original reference [39], this exponent has been incorrectly typed as $-11\ m\,s^{-2}$.

Fig. 2.4 The trend of ηD_T against M for PS in two different solvents, viz., cyclooctane and MEK. The trends are based on the experimental data reported in [53]. Figure modified from [53]

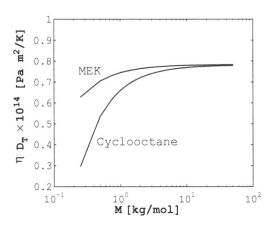

2.3.2   Liquid Polymer Mixtures

Unlike alkanes, theoretical considerations for polymers reveal that viscosity of the solvent plays a more dominant role in the thermodiffusion of polymer solutions. For instance, if one looks at the thermodiffusion in solutions of polystyrene (PS) in toluene [46], it is found that for a molar mass in the range of $5 - 4,000$ kg mol^{-1}, D_T is independent of the molar mass for almost any concentration. In fact, the thermodiffusion coefficient is purely dictated by the microscopic local viscosity, η_{eff}, and one can write a very simple expression for D_T and S_T in such mixtures as

$$D_T = \frac{\Delta_T}{\eta_{\text{eff}}}, \tag{2.123a}$$

$$S_T = \frac{6\pi R_h^a \Delta_T}{(1 - \phi)^2 \kappa_B T}, \tag{2.123b}$$

where Δ_T (Pa m^2 K^{-1}) is a constant that depends upon the polymer. R_h^a and ϕ are the apparent hydrodynamic radius of the polymer and volume fraction of the polymer, respectively. Further, in a very dilute solution, $\eta_{\text{eff}} = \eta_s$, the viscosity of the solvent.

In case of oligomers and short polymer chains, D_T increases with the molecular weight of the polymer and eventually reaches a steady state value for very large M. This trend is seen in Fig. 2.4, where the variation of the product ηD_T with the molecular weight is plotted for the solution of PS in two solvents, viz., cyclooctane and methyl ethyl ketone (MEK). In fact, an analysis of PS in various solvents reveals that the thermodiffusion coefficient can be represented as [53]

$$D_T = \frac{\Delta_T}{\eta} - \frac{a}{M^\alpha}, \tag{2.124}$$

where η is the viscosity of the solvent and a (kg m^2 s^{-1} K^{-1} mol^{-1}) is a amplitude factor that depends upon the solution. From the above equation it is clear that for small values of M, there is an impact of M on D_T. However, for large values, i.e., when the Kuhn segments are more than 1 kg mol^{-1}, the second term becomes irrelevant and the equation looks like (2.123a). More precisely, it is seen that the thermodiffusion coefficient can simply be written as

$$D_T \approx 0.6 \times 10^{-14} \times \eta^{-1}. \tag{2.125}$$

Two other correlations are applicable for the dilute polymer solutions. The first one is proposed by Khazanovich [37] and the other one is by Semenov and Schimpf [48]. According Khazanovich [37], the thermodiffusion coefficient can be written as

$$D_T = \frac{D_b E_{A,s}}{RT^2}, \tag{2.126}$$

where D_b is the self diffusion coefficient of the bead, R is the gas constant, and $E_{A,S}$ is the activation energy. D_b can be calculated using the hydrodynamic radius, R_b, and the solvent viscosity, η_s, as

$$D_b = \frac{\kappa_B T}{6\pi \eta_s R_b}. \tag{2.127}$$

The activation energy can be expressed in terms of the viscosity data of the solvent as

$$E_{A,s} = -RT^2 \frac{d\ln \eta_s}{dT}. \tag{2.128}$$

Since (2.126) makes use of an activation energy principle, one can argue that this expression is more appropriate in the thermodynamic models considered in Sect. 2.1. Nevertheless, due to the fact that this is a much simpler expression requiring only the viscosity data for calculating the derivative in (2.128), we present it in this section.

According to the correlation presented by Semenov and Shimpf [48], for a diminishing concentration of the polymer in a solution, the thermodiffusion coefficient can be estimated as

$$D_T = \frac{8\beta_s \sqrt{A_p A_s} r_m^2}{27 v_s \eta_s} \left(1 - \sqrt{\frac{A_s}{A_p}} \right), \tag{2.129}$$

where β_s is the cubic thermal expansion coefficient of the solvent and r_m is the radius of the monomer. A_p and A_s are the Hamaker constants of the monomer and the solvent, respectively. v_s is the volume occupied by each solvent molecule and is calculated as

$$v_s = \frac{M_s}{\rho_s N_a}, \tag{2.130}$$

where N_a is the Avogadro number.

There is also a simple equation proposed by Brenner [5] following a kinematic analysis which presents the thermodiffusion coefficient as

$$D_T = \lambda \beta_s D_s^s,$$ (2.131)

where λ is a non-ideality factor that is of $O(1)$ and is equal to one for ideal gas mixtures, and D_s^s is the self-diffusion coefficient of the solvent.

Finally, if one considers the temperature effects on the Soret effect, then it is found that the Soret coefficient varies with the mixture temperature via a simple correlation as

$$S_T = S_T^\infty \left[1 - \exp\left(\frac{T^* - T}{T_0} \right) \right],$$ (2.132)

where T^* is the temperature at which the temperature switching occurs and T_0 is some reference temperature in Kelvin. It must be noted that since the numerator of the exponent is a difference between two temperatures, the choice of Celsius or Kelvin should not matter.

2.3.3   Colloidal Mixtures

Thermodiffusion is studied as thermophoresis in colloidal mixtures. The objective in these studies is to determine the separation behavior of the suspended particles in the presence of a temperature gradient in the mixture. Currently, a good physical understanding of thermophoresis is still considerably poor and the direction of separation is also quite unpredictable. While the dispersed particles mostly move to the cold side, there is some experimental evidence showing an opposite result as well. Nevertheless, our current understanding of thermophoresis has produced some simple correlations for the process.

Temperature effects on thermophoresis in colloidal solutions can still be represented by (2.132). In fact, several macromolecular and colloidal systems, including mixtures of proteins and polypeptides, follow this equation. The effect of the three parameters, viz., S_T^∞, T^* and T_0, on the evolution of this equation is illustrated in Figs. 2.5–2.7. As seen in Fig. 2.5, the amplitude factor, S_T^∞, dictates the pace at which the Soret coefficient changes with temperature. Small values of S_T^∞ implies a relatively smaller value of dS_T/dT, and also a lower final steady state value of S_T.

T^* has an effect of shifting the null point in the curve. More precisely, a large value of T^* implies a larger temperature at which a change in the sign of S_T occurs. Nevertheless, in view of the fact that S_T^∞ is constant in these curves, all three lines in Fig. 2.6 converge to the same horizontal asymptote of $S_T^\infty = 0.01$ K^{-1}.

Finally, the contribution of the reference temperature, T_0, is similar to the impact of S_T^∞, i.e., it somewhat controls the pace at which the Soret coefficient changes with temperature. At smaller values of T_0, the value of dS_T/dT is large and S_T reaches the final steady state value of S_T^∞ at a much earlier operating temperature.

Fig. 2.5 The effect of S_T^∞ on the evolution of (2.132). In all three curves shown in the figure $T^* = 15°C$ and $T_0 = 20$ K

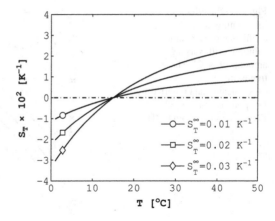

Fig. 2.6 The effect of T^* on the evolution of (2.132). In all three curves shown in the figure $S_T^\infty = 0.01$ K^{-1} and $T_0 = 20$ K

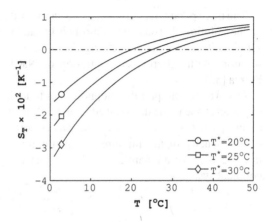

Fig. 2.7 The effect of T_0 on the evolution of (2.132). In all three curves shown in the figure $S_T^\infty = 0.01$ K^{-1} and $T^* = 15$ K

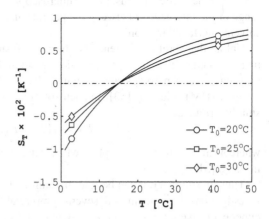

Fig. 2.8 The effect of temperature on the Soret coefficient of lysozome solutions with varying amounts of NaCl. These trends are on the experimental data reported by Iacopini and Piazza [32]. Figure modified from [32]

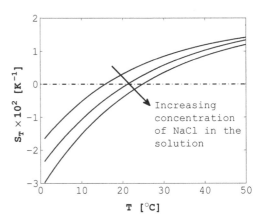

The three parameters just discussed can vary due to a change in the ionic strength of the solution that can be controlled by adding salt, NaCl for instance, in the solution. This is illustrated in Fig. 2.8 where (2.132) is plotted for the lysozome solutions with different concentration of NaCl as investigated by Iacopini and Piazza [32]

One way to interpret thermophoresis is to look at it as a consequence of the drift of particles due to the unbalanced interfacial stresses in the immediate neighborhood of the particle surface. These unbalanced stresses arise due to the temperature inhomogeneity in the mixture. A quantitative consideration of this results in a microscopic hydrodynamic model [33] that presents the Soret coefficient in the colloidal system as

$$S_T = \frac{4\pi r}{\kappa_B T} \frac{\partial (l\gamma)}{\partial T}, \tag{2.133}$$

where r is the particle radius, l is the microscopic length scale that depends upon the range of the interactions between the particle and the solvent, and γ is the particle-solvent interfacial tension.

Unlike the Soret coefficient, the thermodiffusion coefficient varies linearly for many systems including polypeptides, sodium polystyrene sulfonate (NaPSS), polystyrene latex particles, and sodium dodecyl sulfate (SDS)-β-dodecyl-maltoside micellar (DM) solutions [33]. In such solutions, the thermodiffusion coefficient is simply written as

$$D_T = A (T^* - T), \tag{2.134}$$

where A (m^2 s^{-1} K^{-2}) is a amplitude factor that depends upon the type of the system investigated.

Thus, in summary, it can be said that numerous thermodiffusion/thermophoresis investigations have identified several empirical correlations that are prescribed either exclusively for a particular system or for a class of liquid mixtures. Although not complete, with little effort in determining the constants and other easily accessible parameters, these correlations can be conveniently employed to study a variety systems. A primary use of these correlations would be in quickly planning new experiments. In any case, with availability of new experimental data these correlation must be verified and if needed upgraded/corrected.

2.4   Method 3: Neurocomputing Models to Study Thermodiffusion

2.4.1   What are Neural Networks?

Artificial Neural Networks (ANN) or simply *neural networks* (NN) are an attempt to combine the principles of associative approaches and mathematical modeling to develop a *engineering system* that can emulate the human brain. The primary application of such a system would be in studying difficult engineering problems that lack accurate mathematical modeling. Needless to say, before attempting to outline the theoretical framework of such a computational tool, it is pertinent to first introduce some basic neurobiology to understand how the human brain works.

Human brain consists of nearly 100 billion nerve cells that are called *neurons*. A highly simplified example of the functioning of the neurons and the human brain in general is depicted in Fig. 2.9. Essentially, neurons communicate via electrical pulses that are generated via small voltage spikes of the cell walls. The information passes through the *axon* and reaches the other neurons via the *synapses*. Upon receiving these signals, the information is processed and if a sufficient threshold is reached then the recipient neuron will generate a voltage impulse in response. Once again, this is transmitted to the other neurons via the axon. This flow of information/signal is indicated by the arrows in Fig. 2.9.

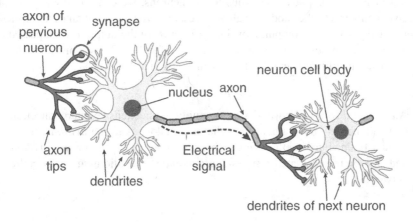

Fig. 2.9 Functioning of a neuron in a simplified format

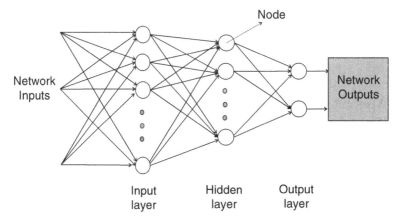

Fig. 2.10 Schematic of an artificial neural network where several layers of nodes are connected

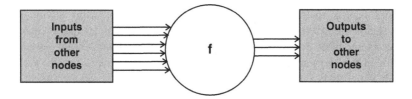

Fig. 2.11 The functionality of a node in an artificial neural network

In solving difficult engineering problems where the response or the output of a system depends upon complex interactions between several parameters that are either known or unknown, this neurobiological principle can be employed to develop what is known as an *Artificial Neural Network*. A schematic of ANN is shown in Fig. 2.10. Thus, analogous to the neurons, we can employ *nodes* that receive inputs from other nodes, evaluate a response function that takes all of these inputs, and calculate an output. A weighted fraction of the output is then passed to the other nodes. A design of a node is depicted in Fig. 2.11. Thus, a more structured definition of an artificial neural network can be given as follows:

An *Artificial Neural Network* is an assembly of an extensive interconnection of fundamental units called *nodes* that are analogous to the *neurons* in the human brain. Further, these *nodes* process a set of *inputs* or *information* and pass on the evaluated *response/output/data* to the adjoining *nodes* in the network.

2.4.2 Why Artificial Neural Networks?

In several instances, some of the physics can be neglected to simplify the mathematical formulation of the problem without compromising the validity or even the accuracy of the solution. However, there are several applications where this simplification is not acceptable. Much worse, despite extensive research, several problems cannot be completely formulated due to our limited understanding of the underlying physics. As a result, a precise mathematical modeling of the entire process is extremely challenging in many engineering problems.

In such cases, when the complexity of the problem at hand cannot be solved exactly via existing mathematical models, the principles of artificial neural networks come in handy as they are applicable to problems where a number of parameters influencing the outcome of a system interact in a complex manner. Additionally, as will be discussed in the subsequent sections, it is also possible to include the effect of the parameters that are either unknown or neglected in determining the output of the system using an incomplete set of input parameters.

In the literature, ANN has been employed by researchers to study a variety of problems such as the mechanical [28, 31], electrical [6, 49], and thermophysical properties of the materials [38]. It has also been used to study tribological properties, viz., wear rate and the friction coefficient of fiber composites [57], corrosion studies [40, 47], and surface texture studies [4, 45].

Like the above problems, thermodiffusion is also an equally intricate phenomenon in which despite nearly a century of research, there is still a lack of a single unified theory that can explain the observations at a molecular or microscopic level. One of the main issues is the complexity of the underlying inter-particle and intra-particle interactions, as well as external forces that can affect the liquid at the molecular level. The application of ANN to study thermodiffusion is therefore pertinent.

2.4.3 Theoretical Formalisms

In this section, a detailed formalism of an artificial neural network that can be used to study thermodiffusion problems is presented. As mentioned earlier, ANN combines the principles of associative thinking with the precision of mathematical modeling. Development of such a neural network model involves the following steps, each of which will be discussed in detail:

1. Choosing a network topology
2. Database generation
3. Neural network training and testing
4. Neural network validation

Fig. 2.12 Examples of some typical neural network topologies. In each case, the inputs enter the neural network from the left through the first layer. The last layer produces the output of the neural network. Each layer can have one or more nodes

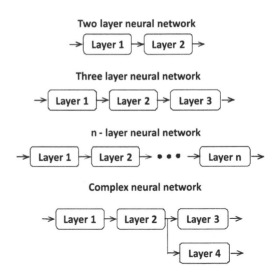

2.4.3.1 Network Topology

Network topology refers to the structure of the network. Specifically, choosing a network topology involves deciding the number of nodes to be employed in each layer of the neural network, the number of such layers and the way in which the layers are interconnected. Some of the suggested topologies for solving problems related to physics are shown in Fig. 2.12. Any of these topologies can be employed to study thermodiffusion.

As seen in this figure, several complicated topologies are possible. Two points must be noted about these topologies:

1. A layer can have an arbitrary number of nodes, i.e., the number of nodes in each layer can be different. A prior study is essential in determining the number of layers on the neural network as well as the number of nodes to be employed in each layer.
2. Usually, the network connections are such that nodes from each layer is connected to every node in the subsequent layer. All the topologies shown in Fig. 2.12 correspond to a *feed forward neural network*. As the name indicates, the inputs enter the neural network through the first layer on the left and propagate from one layer through the next in one direction (right). The output of the last layer represents the output of the neural network. There are other possible interconnections, but they are out of the current scope. The readers can refer to dedicated works on neural networks that describe the other network implementations in greater depth [24, 27].

To determine a neural network topology that is capable of predicting, say for instance, the Soret coefficient, thermodiffusion coefficient or even the mixture properties such as density and viscosity, the the first step is to finalize the number of

Fig. 2.13 The mean square
error analysis to determine
the number of nodes in an
internal layer of the neural
network

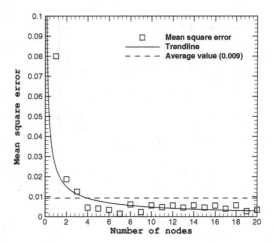

layers in a network. To this end, a three layer network can be employed since it is
sufficient to solve most engineering problems. Next, the number of nodes in these
layers is determined through the analysis of the neural network. More precisely, the
number of nodes are progressively increased until the performance of the neural
network stabilizes, i.e., for a given set of input data, the network predictions are
compared with the experimental data. The mean square error of all the data are
studied by plotting them in a graph as shown in Fig. 2.13. In this example figure, it is
evident that increasing the number of nodes beyond 5 does not produce a noticeable
improvement in the performance of the neural network. Hence, using five nodes in
this layer would suffice.

An appropriate topology is important since the number of network layers and the
number of nodes in each layer determine the ability of the network to capture the
important thermodiffusion trends in the mixtures. It must be noted that a network
with many nodes can result in an over-fitting of the data. On the other hand a network
with too few nodes may not be able to predict the thermodiffusion trends accurately.
This is illustrated in Fig. 2.14 where the data (with some errors) more or less obey
a linear trend. However, with excessive nodes, the error in the data is also modeled
by the neural network and it predicts a polynomial behavior of the data.

As mentioned earlier, the input parameters are fed to the neural network through
the nodes at the extreme left of the network. Each node processes this information,
produces an output, and transfers it to every node in the subsequent layer. In this
way, information propagates through the network and the outputs (corresponding to
the modeled parameters) are obtained from the last layer of nodes in the network.
Mathematically, the output, u, of the ith node in the kth layer of the network is [27]

$$u_i^{(k)} = f^{(k)} \left(b_i^{(k)} + \sum_{j=1}^{J} u_j^{(k-1)} w_{ji}^{(k)} \right), \tag{2.135}$$

Fig. 2.14 An illustration of the over-fitting of the experimental data by the neural network

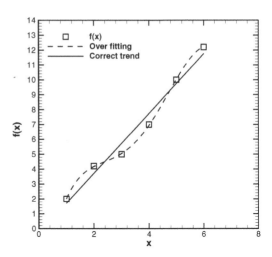

where w are *weight coefficients* that scale the inputs to a node and J is the number of nodes in the layer $(k-1)$. b is called the *bias* term that is included in the above expression to account for the following:

1. Parameters that have been missed in the input parameter set which impact the process and thereby the output variables.
2. Other unknown parameters that might be contributing to the physics of the problem.

In (2.135), f is a *transformation function* that modulates the sum of the bias and weighted inputs. While different types of transformation functions can be employed, two popular functions are

$$f(\zeta) = \tanh(\zeta) = \frac{1 - e^{-2\zeta}}{1 + e^{-2\zeta}}, \tag{2.136a}$$

$$f(\zeta) = \frac{1}{1 + e^{-\zeta}}. \tag{2.136b}$$

2.4.3.2 Database Generation

In this step, experimental data of the quantities of interest, viz., Soret coefficient, thermodiffusion coefficient, thermodiffusion factor, mixture density, mixture viscosity, etc. are collected for various input parameter combinations. In typical thermodiffusion studies, the input parameter set would include the mole fraction of the participating components and pure component properties like the molecular weight, density, and viscosity.

Following the compilation of the database, before using it with an artificial neural network, *standardization* and *normalization* of the database are common practice.

These are necessary and are usually done to shield the neural network from the influence of the absolute value of a particular parameter. Standardization is done by adjusting the input parameters of the database such that each parameter set in the database has a mean of zero and a standard deviation of one. This can be done via the following equation

$$p_{i,:} = \frac{(p'_{i,:} - \bar{p}_i)}{\sigma_i}, \tag{2.137}$$

where $p_{i,:}$ are the values of the ith standardized input parameter, and $\bar{p}_i$ and σ_i are the mean and standard deviations, respectively, of the ith input parameter. $p'_{i,:}$ are the actual values of the parameters in the respective experiments.

Following this, the range of each parameter is normalized such that $p_i \in \{-1, +1\}$. This can be achieved by applying the following transformation to the standardized database.

$$p_{i,:} = \frac{(p_{i,:} - p_{i,:}^{min})}{(p_{i,:}^{max} - p_{i,:}^{min})}, \tag{2.138}$$

where $p_{i,:}^{min}$ and $p_{i,:}^{max}$ represent the minimum and maximum values, respectively, of the ith parameter in the standardized database.

2.4.3.3   Neural Network Training

As the input data propagates through each node, it gets transformed via a set of *bias* and *weights* values. The agreement between the output values predicted by the neural network and the experimental data (called the *target* values) is directly governed by the accuracy of the set of weights and bias values. To obtain the optimal combination of the bias and weights values that result in a good network, a search algorithm is applied. This process is called *network training* or simply *training* which will culminate in a network that is capable of predicting the target data with the least error.

For training a neural network, the standardized and normalized database is randomly shuffled and split into two parts. Usually, 60–80% of the randomly shuffled database is employed for training and is called the *training data*. The remaining 40–20% of the unused database is saved to evaluate the trained neural network and is called the *validation data*. These data are not seen by the NN while it is being trained.

For the optimal weight determination problem in hand, one could resort to gradient-based search algorithms or heuristic search methods. The former includes Newton's method, Conjugate gradient method, Levenberg–Marquardt backpropagation (LMBP) algorithm, etc. On the other hand, Genetic algorithm-based searches come in the category of heuristic methods.

A gradient-based training algorithm that is well established in the literature is the *LMBP* algorithm [27] that minimizes a mean square error function over a multidimensional parameter space. This algorithm begins with a network initialization in

which a random set of values for weights and bias are chosen as an initial guess.
One could also employ a more systematic initialization method such as the Nguyen
and Widrow's algorithm [42]. Following the initialization of the network weights
and bias values, the algorithm performs the following steps iteratively:

Step 1: Evaluate the response of the network for the training data set and compare
it with the desired experimental data.
Step 2: If the convergence criterion is not reached, update the network weights and
bias values, and go to Step 1.

To update the weights, the LMBP algorithm uses an enhanced version of the
Newton's method that is derived as follows: In the kth iteration of the algorithm, a
Newton's update for a node that receives N input parameters and that are scaled by
a weight vector $\mathbf{w} \in \Re^{N \times 1}$, can be written as

$$\mathbf{w}(k+1) = \mathbf{w}(k) - \mathbf{H}_k^{-1}\mathbf{g}_k, \tag{2.139}$$

where the Hessian, $\mathbf{H}_k$, is a real valued matrix, i.e., $\mathbf{H}_k \in \Re^{N \times N}$ and $\mathbf{g}_k \in \Re^{N \times 1}$ is
the gradient of mean square error, E. The subscript, k, in $\mathbf{H}_k$ and $\mathbf{g}_k$ indicates that
these quantities are evaluated at $\mathbf{w} = \mathbf{w}(k)$. In other words,

$$\mathbf{H}_k = \nabla^2 E \big|_{\mathbf{w}=\mathbf{w}(k)}, \tag{2.140}$$

$$\mathbf{g}_k = \nabla E \big|_{\mathbf{w}=\mathbf{w}(k)} .$$

In the above equations, E is a measure of the error and is expressed as

$$E = \frac{1}{2\zeta} \sum_{q=1}^{\zeta} (\mathbf{t}_q - \mathbf{y}_q)^{\mathrm{T}} (\mathbf{t}_q - \mathbf{y}_q) = \frac{1}{2\zeta} \sum_{q=1}^{\zeta} \sum_{j=1}^{J} (t_{q,j} - y_{q,j})^2, \tag{2.141}$$

where ζ is the total number of training vectors that are applied in each iteration. The
target vector, $\mathbf{t}$, has J outputs and the vector $\mathbf{y}$ consists of the corresponding output
values predicted by the network.

In the expression of the Hessian, the $\nabla^2 E$ term can be written in terms of the
Jacobian, $\mathbf{J}$ as

$$\nabla^2 E = \mathbf{J}^{\mathrm{T}}\mathbf{J} + \boldsymbol{\Phi}. \tag{2.142}$$

$\mathbf{J}$, in the above equation, has the elements

$$\mathbf{J} = \frac{1}{\sqrt{\zeta}}
\begin{bmatrix}
\frac{\partial(t_1-y_1)}{\partial w_1} & \frac{\partial(t_1-y_1)}{\partial w_2} & \cdots & \frac{\partial(t_1-y_1)}{\partial w_N} \\
\frac{\partial(t_2-y_2)}{\partial w_1} & \frac{\partial(t_2-y_2)}{\partial w_2} & \cdots & \frac{\partial(t_2-y_2)}{\partial w_N} \\
\vdots & \vdots & \ddots & \vdots \\
\frac{\partial(t_{\zeta J}-y_{\zeta J})}{\partial w_1} & \frac{\partial(t_{\zeta J}-y_{\zeta J})}{\partial w_2} & \cdots & \frac{\partial(t_{\zeta J}-y_{\zeta J})}{\partial w_N}
\end{bmatrix}. \tag{2.143}$$

Table 2.2 The key steps of the Levenberg–Marquardt backpropagation algorithm [27], reproduced from [52]

Step 1: Initialize network weights and biases.

while $\mathbf{w} \neq \mathbf{w}^*$

for $j = 1, 2, \cdots, P$

Step 2: Calculate the network output for a particular data vector.

Step 3: Calculate the jth row of the Jacobian matrix in (2.143).

end for loop

Step 4: Update network weights using (2.149) and check if $\mathbf{w} = \mathbf{w}^*$.

end while loop

Also, Φ in (2.142) is

$$\Phi = \sum_{i=1}^{P} e_i \nabla^2 e_i, \tag{2.144}$$

where $e_i = (t_i - y_i)/\sqrt{\zeta}$ and $P = \zeta J$. Close to the convergence criterion, $\mathbf{w}$ approaches the optimal value of $\mathbf{w}^*$, $e_i \to 0$ and $\Phi \to 0$. In other words, we can approximate the Hessian as

$$\mathbf{H} \approx \mathbf{J}^T \mathbf{J}. \tag{2.145}$$

Further, the gradient can be written in terms of the Jacobian as

$$\mathbf{g} = \mathbf{J}^T \mathbf{e}, \tag{2.146}$$

and we can rewrite (2.139) as

$$\mathbf{w}(k+1) = \mathbf{w}(k) - \left[\mathbf{J}_k^T \mathbf{J}_k\right]^{-1} \mathbf{J}_k^T \mathbf{e}_k. \tag{2.147}$$

Now, it must be noted that in the above equation, the inversion of the Hessian can be an ill-conditioned problem. To overcome this drawback one can introduce a small perturbation to the Hessian and write it as

$$\mathbf{H} \approx \mathbf{J}^T \mathbf{J} + \xi \mathbf{I}, \tag{2.148}$$

where $\mathbf{I} \in \mathfrak{R}^{N \times N}$ is an identity matrix and ξ is a small scalar called the *learning parameter*. Introducing this approximation of Hessian in (2.147), we obtain the Levenberg–Marquardt update as

$$\mathbf{w}(k+1) = \mathbf{w}(k) - \left[\mathbf{J}_k^T \mathbf{J}_k + \xi_k \mathbf{I}\right]^{-1} \mathbf{J}_k^T \mathbf{e}_k. \tag{2.149}$$

Theoretically, convergence must be obtained after a large number of iterations. More precisely, as $k \to \infty$, $\mathbf{e}_k \to 0$ and $\mathbf{w} \to \mathbf{w}^*$. In other words, $\mathbf{w}$ converges to the most appropriate weight matrix $\mathbf{w}^*$ that enables the network to predict the target values with minimal error. The key steps of the LMBP algorithm are outlined in Table 2.2.

2.4.3.4 Termination Criterion of the Algorithm

Keeping in mind the numerics of the implementation of the algorithm, the optimal weight determination procedure described above is coupled with a set of termination criterion that are checked at each iteration of the training process. These criteria are as follows:

1. The training is terminated if the mean square error is less than tol_{mse}. The choice of this tolerance value is decided after some numerical experimentation. This is because a very small error tolerance can result in an over-fitting of the experimental data, and with a large value of tol_{mse} there is a risk of developing a neural network model with poor prediction capabilities.
2. During the course of network training, the updated network in each iteration is tested with respect to a small data set. These *test* data are a subset of the training data and are *not* taken from the validation data set. Now, in evaluating these test data, if the network performance does not improve for tol_t successive iterations, then it is assumed that the network training is complete and the training process is terminated.
3. Since the LMBP algorithm is a gradient-based search algorithm, the zero gradient point corresponds to an optimal value. However, an absolute zero gradient is perhaps too stringent a condition to implement. Hence, the training is usually stopped when the gradient in (2.146) is sufficiently small. In other words, the training is terminated if the gradient is less than tol_{grad}.
4. Irrespective of the above conditions for ending the optimization process, the algorithm is terminated if the number of iteration reaches a specified upper limit, i.e., $P \geq P_{max}$. While the previous three termination criteria indicate some sort of convergence to an optimum, this criterion indicates a poor convergence and must be generally avoided. For this, a large value of P_{max} must be specified. In the thermodiffusion studies, it has been found that the use of $P_{max} \approx O(10^5)$ is sufficient to ensure that the optimization process terminates in one of the first three conditions, i.e., an optimal set of weights is found.

2.4.3.5 Neural Network Validation

This is the final step in which the generated neural network is tested with some experimental data to ensure that the neural network is well trained. For this step, the part of the database that was not used for the training is generally employed. It is important to ensure that this validation data set is sufficiently diverse. This is to verify the prediction abilities of the neural network for the entire range of the parameters in the input data for which the neural network has been trained.

In addition to the testing of the neural network with respect to the experimental data values, it is also essential to conduct studies on its abilities to predict the important trends. For instance, it should be able to predict the effect of the relative molecular weights in a binary n-alkane mixture, for instance, where we know that the thermodiffusion process is more pronounced if the relative molecular weights are very large. Also, in some mixtures, a change in the direction of separation has been observed experimentally. A good neural network should be able to predict this direction switch. The application of neural network approach, discussed in this section, to binary liquid mixtures is taken up in detail in Chap. 6. There the ability of the neural network to predict these trends is considered more elaborately.

References

1. Artola PA, Rousseau B, Galliero G (2008) A new model for thermal diffusion: kinetic approach. J Am Chem Soc 130:10,963–10,969
2. Blanco P, Bou-Ali M, Platten JK, Urteaga P, Madariaga JA, Santamaria C (2008) Determination of thermal diffusion coefficient in equimolar n-alkane mixtures: empirical correlations. J Chem Phys 129:174,504. http://dx.doi.org/10.1063/1.2945901
3. Blanco P, Bou-Ali M, Platten JK, de Mezquia DA, Madariaga JA, Santamaria C (2010) Thermodiffusion coefficients of binary and ternary hydrocarbon mixtures. J Chem Phys 132:114,506. http://dx.doi.org/10.1063/1.3354114
4. Brahme A, Winning M, Raabe D (2009) Prediction of cold rolling texture of steels using an artificial neural network. Comput Mater Sci 46(4):800–804. URL http://www.sciencedirect.com/science/article/pii/S0927025609001839 DOI: 10.1016/j.commatsci.2009.04.014
5. Brenner H (2006) Elementary kinematical model of thermal diffusion in liquids and gases. Phys Rev E 74:036,306
6. Cai K, Xia J, Li L, Gui Z (2005) Analysis of the electrical properties of PZT by a BP artificial neural network. Comput Mater Sci 34:166–172
7. Debuschewitz C, Köhler W (2001) Molecular origin of thermal diffusion in benzene+cyclohexane mixtures. Phys Rev Lett 87:055,901
8. Denbigh KG (1952) The heat of transport in binary regular solutions. Trans Faraday Soc 48:1–8
9. Dhont JKG (2004) Therodiffusion of interacting colloids. I. A statistical thermodynamics approach. J Chem Phys 120(3):1632–1641
10. Dougherty EL, Drickamer HG (1955a) A theory of thermal diffusion in liquids. J Chem Phys 23(5):295
11. Dougherty EL, Drickamer HG (1955b) Thermal diffusion and molecular motion in liquids. J Phys Chem 59(5):443–449
12. Eslamian M, Saghir MZ (2009) A dynamic thermodiffusion model for binary liquid mixtures. Phys Rev E 80:011,201
13. Eslamian M, Saghir MZ (2009) Microscopic study and modeling of thermodiffusion in binary associating mixtures. Phys Rev E 80:061,201
14. Eslamian M, Saghir MZ (2010a) Dynamic thermodiffusion theory for ternary liquid mixtures. J Non-Equilib Thermodyn 35:51–73
15. Eslamian M, Saghir MZ (2010b) Investigation of the Soret effect in binary, ternary and quaternary hydrocarbon mixtures: new expressions for thermodiffusion factors in quaternary mixtures. Int J Therm Sci 49:2128–2137
16. Eslamian M, Saghir MZ (2011) Estimation of thermodiffusion coefficients in ternary associating mixtures. Can J Chem Eng 9999:1–8
17. Eslamian M, Saghir MZ (2011) Non-equilibrium thermodynamic model fro the estimation of soret coefficient in dilute polymer solutions. Int J Thermophys 32:652–664
18. Eslamian M, Saghir MZ (2012) Modeling of DNA thermophoresis in dilure solutions using the non-equilibrium thermodynamics approach. J Non-Equilib Thermodyn 37:63–76
19. Ghorayeb K, Firoozabadi A (2000) Molecular, pressure, and thermal diffusion in non-ideal multicomponent mixtures. AIChE J 46(5):883–891 http://dx.doi.org/10.1002/aic.690460503

20. Glasstone S, Laidler KJ, Eyring H (1941) The theory of rate processes. The kinetics of chemical reactions, viscosity, diffusion and electrochemical phenomena. McGraw-Hill, New York
21. de Groot SR, Mazur P (1984) Non-equilibrium thermodynamics. Dover, New York
22. Gross J, Sadowski G (2001) Perturbed-chain saft: an equation of state based on a perturbation theory for chain molecules. Ind Eng Chem Res 40(4):1244–1260
23. Gross J, Sadowski G (2002) Modeling polymer systems using the perturbed-chain statistical associating fluid theory equation of state. Ind Eng Chem Res 41:1084–1093
24. Gurney K (1997) An introduction to artificial neural networks, 1st edn. UCL, Taylor & Francis Group, 1 Gunpowder Square, Londo EC4A 3DE
25. Guy AG (1986) Prediction of thermal diffusion in binary mixtures of nonelectrolyte liquids by the use of nonequilibrium thermodynamics. Int J Thermophys 7:563–572
26. Haase R (1969) Thermodynamics of irreversible processes. Addison-Wesley, Reading
27. Ham FM, Kostanic I (2001) Principles of neurocomputing for science and engineering, 2nd edn. McGraw Hill, New York
28. Hancheng Q, Bocai X, Shangzheng L, Fagen W (2002) Fuzzy neural network modeling of material properties. J Mater Process Technol 28:196–200
29. Hartmann S, Königer A, Köhler W (2008) Isotope and isomer effect in thermal diffusion of binary liquid mixtures. In: Köhler W, Wiegand S, Dhont JKG (eds) Thermal nonequilibrium. Proceedings of the eighth international meeting on thermodiffusion. Forschungszentrum Jölich GmbH, Bonn, pp 35–41
30. Hiemenz PC, Lodge TP (2007) Polymer chemistry, 2nd edn. CRC, Boca Raton
31. Hwang RC, Chen YJ, Huang HC (2010) Artificial intelligent analyzer for mechanical properties of rolled steel bar by using neural networks. Expert Syst Appl 37(4):3136–3139. URLhttp://www.sciencedirect.com/science/article/pii/S0957417409008501 DOI:10.1016/j.eswa.2009.09.069
32. Iacopini S, Piazza R (2003) Thermophoresis in protein Solutions. Europhys Lett 63(2): 247–253
33. Iacopini S, Rusconi R, Piazza R (2006) The "macromolecular tourist": universal temperature dependence of thermal diffusion in aqueous colloidal suspensions. Eur Phys J E 19:59–67
34. Jhaverl BS, Youngren GK (1988) Three-parameter modification of the Peng-Robinson equation of state to improve volumetric predictions. SPE Reserv Eng 3(3):1033–1040
35. Kempers LJTM (1989) A thermodynamic theory of the soret effect in a multicomponent liquid. J Chem Phys 90:6541–6548
36. Kempers LJTM (2001) A comprehensive thermodynamic theory of the soret effect in a multicomponent gas, liquid, or solid. J Chem Phys 115:6330–6341
37. Khazanovich TN (1967) On theory of thermal diffusion in dilute polymer solutions. J Polym Sci C: Polym Symp 16:2463–2468
38. Laugier S, Richon D (2003) Use of artificial neural networks for calculating derived thermodynamic quantities from volumetric property data. Fluid Phase Equil 210(2):247–255. URL http://www.sciencedirect.com/science/article/pii/S0378381203001729 DOI: 10.1016/S0378-3812(03)00172-9
39. Madariaga JA, Santamaria C, Bou-Ali M, Urteaga P, De Mezquia DA (2010) Measurement of thermodiffusion coefficient in n-alkane binary mixtures: composition dependence. J Phys Chem B 114:6937–6942. http://dx.doi.org/10.1021/jp910823c
40. Martin Ó, De Tiedra P, López M (2010) Artificial neural networks for pitting potential prediction of resistance spot welding joints of AISI 304 austenitic stainless steel. Corros Sci 52:2937–2402
41. Morozov KI (2009) Soret effect in molecular mixtures. Phys Rev E 79:031,204
42. Nguyen D, Widrow B (1990) Improving the learning speed of 2-layer neural network by chosing initial values of the adaptive weights. Proc Int Joint Conf Neural Networks 3:21–26
43. Peng D, Robinson DB (1976) A new two-constant equation of state. Ind Eng Chem Fundam 15(1):59–64. http://dx.doi.org/10.1021/i160057a011
44. Prigogine I, de Brouckere L, Amand R (1950) Recherches sur la thermodiffusion en phase liquide: (premiere communication). Physica 16(7–8):577–598

References   55

45. Rashidi A, Hayati M, Rezaei A (2011) Prediction of the relative texture coefficient of nanocrystalline nickel coatings using artificial neural networks. Solid State Sciences 13:1589–1593
46. Rauch J, Köhler W (2003) Collective and thermal diffusion in dilute, semidilute, and concentrated solutions of polystyrene in toluene. J Chem Phys 119:11,977
47. Rolich T, Rezić I, Ćurković L (2010) Estimation of steel guitar strings corrosion by artificial neural network. Corros Sci 52:996–1002
48. Semenov S, Schimpf M (2004) Thermophoresis of dissolved molecules and polymers: consideration of the temperature-induced macroscopic pressure gradient. Phys Rev E 69(2):011,201
49. Seo DC, Lee JJ (1999) Damage detection and CFRP laminates using electrical resistance measurement and neural network. Compos Struct 47:525–530
50. Shukla K, Firoozabadi A (1998) A new model of thermal diffusion coefficients in binary hydrocarbon mixtures. Ind Eng Chem Res 37(8):3331–3342
51. Song S, Peng C (2008) Viscosities of binary and ternary mixtures of water, alcohol, acetone, and hexane. J Disp Sci Tech 29(10):1367–1372
52. Srinivasan S, Saghir MZ (2012) Modeling of thermotransport phenomenon in metal alloys using artificial neural networks. Appl Math Modell. DOI:10.1016/j.apm.2012.06.018
53. Stadelmaier D, Köhler W (2009) Thermal diffusion of dilute polymer solutions: the role of chain flexibility and the effective segment size. Macromolecules 42:9147–9152
54. Tichacek LJ, Kmak WS, Drickamer HG (1956) Thermal diffusion in liquids; the effect of nonideality and association. J Phys Chem 60:660–665
55. Wittko G, Köhler W (2005) Universal isotope effect in thermal diffusion of mixtures containing cyclohexane and cyclohexane-d12. J Chem Phys 123:14,506
56. Wittko G, Köhler W (2007) On the temperature dependence of thermal diffusion of liquid mixtures. Europhys Lett 78:46,007
57. Zhang Z, Friedrich K, Veiten K (2002) Prediction on tribological properties of short fibre composites using artificial neural networks. Wear 252:668–675

Chapter 3
Application of LNET Models to Study Thermodiffusion

Abstract Linear nonequilibrium thermodynamic formulations to study thermodiffusion in liquid mixtures were introduced in Sect. 2.2 of Chap. 2. In this chapter, application of some of these formulations to thermodiffusion in different types of fluid mixtures and the outcomes are presented.

3.1 Liquid Hydrocarbon Mixtures

In this section, the application of two thermodiffusion models to hydrocarbon mixtures will be illustrated, viz., the Firoozabadi model and the Kempers model. These models are described in Sect. 2.2 of Chap. 2. In the ensuing paragraphs, the mixtures considered are at atmospheric pressure.

3.1.1 Binary Hydrocarbon Mixtures

Thermodiffusion in six equimass binary hydrocarbon mixtures at atmospheric pressure are summarized in Fig. 3.1. The temperature conditions are as reported by Pan et al. [4]. As seen in this figure, the overall performance of both models is very good in the binary mixtures. However, between the two models themselves, there are disagreements and depending upon the mixture, one model outperforms the other. These disagreements are mainly because of the way in which the net heat of transport is modeled. While the Firoozabadi model represents Q^* in terms of the partial molar energy of the components, the Kempers model formulates Q^* using the partial molar enthalpy.

S. Srinivasan and M.Z. Saghir, *Thermodiffusion in Multicomponent Mixtures*,
SpringerBriefs in Applied Sciences and Technology, DOI 10.1007/978-1-4614-5599-8_3,
© Springer Science+Business Media, LLC 2013

Fig. 3.1 Thermodiffusion factor of several binary hydrocarbon mixtures at 1 atm and a 50–50 wt% composition. The experimental data and temperature conditions are given in [4]

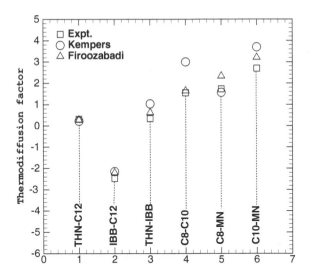

3.1.2 Ternary Hydrocarbon Mixtures

The thermodiffusion coefficients of three ternary hydrocarbon mixtures of nC_{12}–isobutylbenzene (IBB)–tetralin (THN) and nC_8–nC_{10}–methylnaphthalene (MN) evaluated with these two models are shown in Fig. 3.2. For each mixture, all three thermodiffusion coefficients are calculated. Both models work well for the nC_{12}-IBB-THN mixture in which the mass fraction of all three components are $1/3$. However, the Kempers model predicts a wrong sign for the first component in the nC_{12}–IBB–THN mixture. Noting that the thermodiffusion coefficient is very small for this component, an inability of the Kempers model to predict this is perhaps an indicator that it is not very sensitive for very small values of the thermodiffusion coefficients.

In the nC_8–nC_{10}–MN mixture, a similar result is obtained when the mass fraction of each component is $1/3$. However, as the composition changes, there are more discrepancies in the calculations. Specifically, the errors are much larger for the first two thermodiffusion coefficients. In fact, for the second thermodiffusion coefficient of this last mixture, both models predict the wrong sign, i.e., a wrong direction of separation. It must be noted that in applying these models, there might be differences in the values of the calculated coefficients even with the same model. This is because of the use of slightly different values of the model parameters that are not uniform in the literature.

Fig. 3.2 Thermodiffusion coefficient of two ternary hydrocarbon mixtures at 1 atm. The experimental data and temperature conditions are given in [4]

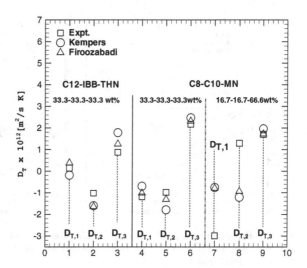

3.2   Liquid Associating Mixtures

Three models are applied to two associating mixtures of various compositions. These are the models of Haase, Firoozabadi, and Eslamian–Saghir, presented in Chap. 2. For the model proposed by Eslamian and Saghir, of the three equations presented by the authors, one corresponding to (2.82) has been evaluated. All three models are coupled with the *PC-SAFT* equation of state, that is well suited for associating mixtures, for ease of comparison. The results of these models for two mixtures, viz., water–ethanol and water–methanol of various mole fractions of water are presented in Fig. 3.3a, b.

As expected, the errors are the largest in the Haase model, whereas the Firoozabadi model is somewhat more accurate. The accuracy of the latter is attributed to the fact that the model depends upon several tuning parameters that can be slightly tweaked to enhance the performance of the model. At present, the most accurate model seems to be the one by Eslamian and Saghir that is at least able to qualitatively predict the sign change in both mixtures in these figures. Specifically, in close agreement with the experiments, the model predicts the sign change at low concentration of the alcohol. Of course, there are still large errors in the accuracy of the magnitude of α in very dilute limits of the compositions. This is because of the complex interactions between the molecules due to additional forces like hydrogen bonding, which is still not very accurately represented in the model.

3.3   Dilute Polymer Mixtures

Application of the thermodiffusion model of Eslamian and Saghir [3] to obtain the thermodiffusion coefficient of poly(methyl methacrylate) (PMMA) in various solvents is summarized in Table 3.1. The overall thermodiffusion trend is well

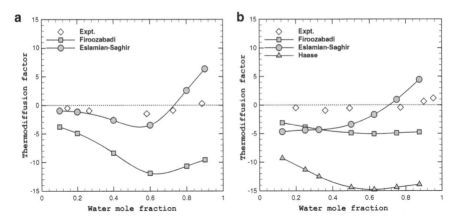

Fig. 3.3 Experimental and calculated thermodiffusion factor of (**a**) water–ethanol mixtures and (**b**) water–methanol mixtures. The experimental data and conditions are the ones reported in [8]. Figures modified from [2]

Table 3.1 Measured and calculated thermodiffusion coefficients of PMMA in various solvents. The experimental data are from [6,7]. Calculated values are from the model of Eslamian and Saghir [3]

Solvent	$D_T^{\text{expt.}} \times 10^{12}$ $[\text{m}^2 \text{ s}^{-1} \text{ K}^{-1}]$	$D_T^{\text{calc.}} \times 10^{12}$ $[\text{m}^2 \text{ s}^{-1} \text{ K}^{-1}]$
Cyclohexanone	3.58 ± 0.77	1.33
THF	12.27 ± 1.05	7.81
Toluene	12 ± 4.97	6.84
Ethyl acetate	11.57 ± 1.87	9
MEK	22.86 ± 1.87	10.05

estimated by the model. A large error is observed when MEK is used as the solvent. Two probable reasons are: (1) the large difference between the molecular weights and densities of PMMA and MEK; (2) in making the calculations, the Mark–Houwink parameter is fixed for all the mixtures, whereas it should be a variable.

The application of the model to calculate D_T of polystyrene in two solvents, viz., ethyl acetate and tetrahydrofuran (THF) at 295 K are shown in Fig. 3.4 for various molecular weights of polystyrene. The experimental data reported in [6, 7] are also shown for comparison. In both mixtures, the model is able to capture the logarithmic trend in the variation of D_T. In the polystyrene–ethyl acetate mixtures, the model predictions are in close agreement with the experimental data.

In the polystyrene–tetrahydrofuran mixtures, the model is able to predict the sign change, as is observed in the experiments. However, the model predictions

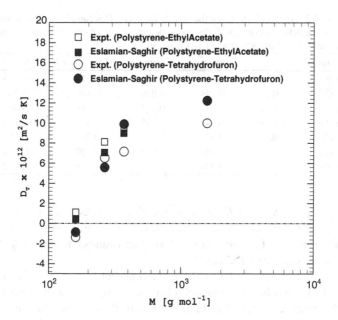

Fig. 3.4 Thermodiffusion coefficient of polystyrene in two different solvents, viz., ethyl acetate and tetrahydrofuran, for various molecular weights of polystyrene. The experimental data and conditions are from [6, 7]

level off much higher than the experimental data, resulting in larger errors at higher molecular weights. This is perhaps due to the role of other physical and chemical properties of the polymer in the thermodiffusion process.

3.4 Molten Metal Mixtures

Over the years, different thermodynamic-based approaches to study thermodiffusion in liquid metals have evolved. In this section we apply the models of Haase, Kempers, and Drickamer to several binary molten metal mixtures for which experimental data are provided by Winter and Drickamer [9]. In the Haase and Kempers models, the ideal gas contribution is neglected for the simplification of the model. In addition to these, a relatively recent model to study thermodiffusion in molten metals presented by Eslamian and Saghir [1] is also considered. This model accounts for the electronic contribution to the thermotransport in the molten metals. This electronic contribution is the mass diffusion due to an internal electric field that is induced as a result of the imposed thermal gradient. The expressions for α_T in this model is:

$$\alpha_T = \frac{E_1^{visc} - E_2^{visc}}{x_1(\partial\mu_1/\partial x_1)} + \frac{-|e|(zS_1 - z_1S)TN}{x_1(\partial\mu_1/\partial x_1)}, \tag{3.1}$$

Table 3.2 α_T calculated using the Haase, Kempers, Drickamer, and Eslamian–Saghir models, compared with the experimental data of equimolar molten metal mixtures studied by Winter and Drickamer [9]

Mixture	Expt.	Haase	Kempers	Drickamer	Eslamian
Sn-Bi	−0.10	−0.01	0.002	−0.518	−0.396
Sn-Cd	−0.35	0.141	0.026	0.140	−0.263
Sn-Zn	−4.10	0.709	0.152	−1.918	−3.085
Sn-Pb	−1.90	−0.146	0.065	−1.331	−1.143
Sn-Pb	−0.83	−0.146	0.061	−0.778	−0.685
Sn-Ga	−0.18	0.101	−0.028	0.789	0.374
Bi-Pb	−1.13	−0.004	0	0.005	−0.072

where e, z, z_1, S, S_1, and N are electron charge, valence of the ions in the mixture, valence of the ions of component 1, thermoelectric power of the mixture, thermoelectric power of component 1 and Avogadro number, respectively. All other notations are as explained in Chap. 2.

Since the other models lack this electronic contribution, for ease of comparison, one can add this contribution the models of Haase, Kempers, and Drickamer to make them look complete. Despite this enhancement of these models, a comparison of these models still points to a superior performance of the formulation by Eslamian and Saghir. It must be noted that there is still large errors in the predictions by these models, and this is primarily due to an inappropriate equation of state. For the calculations shown in Table 3.2, the models have been coupled with perturbed hard-sphere equation of state, described in detail by Eslamian and Saghir [1]. The accuracy can be enhanced further if a good equation of state is formulated for these ionic mixtures.

3.5   Effect of the Equation of State

Two different equations of state were presented in detail in Chap. 2. In this section, the effects of these equations of state on the thermodiffusion calculations are presented. For the calculations, the models of Kempers and Firoozabadi are considered. Both models are coupled with the *v-PR* equation of state and the *PC-SAFT* equation of state. The thermodiffusion coefficients of six binary and three ternary hydrocarbon mixtures are presented in Fig. 3.5a, b.

In both, binary and ternary mixtures, the Firoozabadi model coupled with the *v-PR* equation of state is the most accurate and is able to predict the thermodiffusion coefficients fairly accurately. Further, for this model, the largest error are in the nC_8–nC_{10}–MN mixture with a composition of 16.7–16.7–66.6 wt%. On the other hand, the Kempers model with the *PC-SAFT* equation of state is the most error prone combination to be employed.

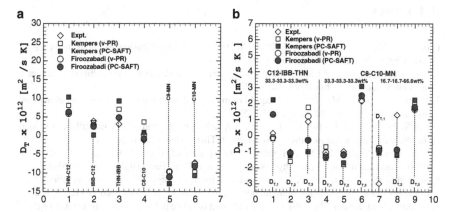

Fig. 3.5 Experimental and calculated thermodiffusion coefficients of (**a**) six binary hydrocarbon mixtures and (**b**) three ternary hydrocarbon mixtures. The experimental data are the ones reported in [4]

The good performance of the Firoozabadi model coupled with *v-PR* equation of state is because of the following: (1) the matching parameters of this model can be tuned to make the model suitable for a certain mixture. Generally, via an initial tuning using some experimental data, the parameters can be adjusted to obtain good results. Subsequently, using these tuning values to study other mixtures does not produce very large errors. (2) the density predictions by the *v-PR* equation of state is very accurate. In fact *PC-SAFT* is notorious in overpredicting the density that has a negative impact on the thermodiffusion coefficients [5]. Also, *PC-SAFT* is primarily designed for associating mixtures. It uses just the partial molar enthalpy in calculating the gradient of chemical potential and eventually underpredicts the value. This means that the thermodiffusion coefficient is inflated in magnitude [5].

In summary, the application of the different thermodiffusion models presented in Chap. 2 to study thermodiffusion in various types of mixtures has been demonstrated in this chapter. Each model performs differently and is good in predicting the thermodiffusion parameters in some mixtures, whereas its performance can be very poor in others. In evaluating the sign change effects in the mixtures, it is seen that some of these models are able to predict the sign change qualitatively. Absolute accuracy of the models for a wide variety of mixtures is still lacking. This is because of the lack of accurate representation of the chemical effects, inter-particle interaction forces such as hydrogen bonding, etc.

Finally, the equation of state is also an important aspect of these models, which goes a long way in determining the accuracy of the models. For instance, a thermodiffusion model coupled with *v-PR* equation of state is more suited for hydrocarbon mixtures than using the *PC-SAFT* equation of state. The latter is more appropriate for associating mixtures.

References

1. Eslamian M, Sabzi F, Saghir MZ (2010) Modeling of thermodiffusion in liquid metal alloys. Phys Chem Chem Phys 12:13,835–13,848
2. Eslamian M, Saghir MZ (2009) Microscopic study and modeling of thermodiffusion in binary associating mixtures. Phys Rev E 80:061,201
3. Eslamian M, Saghir MZ (2010) Nonequilibrium thermodynamic model for soret effect in dilute polymer solutions. Int J Thermophys 32:652–664
4. Pan S, Yan Y, Jaber TJ, Kawaji M, Saghir MZ (2007) Evaluation of thermal diffusion models for ternary hydrocarbon mixtures. J Non-Equilib Thermodyn 32:241–249
5. Srinivasan S, Saghir MZ (2010) Significance of equation of state and viscosity on the thermodiffusion coefficients of a ternary hydrocarbon mixture. High Temp High Pressur 39:65–81
6. Stadelmaier D, Köhler W (2008) From small molecules to high polymers: investigation of the crossover of thermal diffusion in dilute polystyrene solutions. Macromolecules 41:6205–6209
7. Stadelmaier D, Köhler W (2009) Thermal diffusion of dilute polymer solutions: the role of chain flexibility and the effective segment size. Macromolecules 42:9147–9152
8. Tichacek LJ, Kmak WS, Drickamer HG (1956) Thermal diffusion in liquids; the effect of non-ideality and association. J Phys Chem 60:660–665
9. Winter FR, Drickamer HG (1955) Thermal diffusion in liquid metals. J Phys Chem 59(12):1229–1230

Chapter 4
CFD Studies in Thermodiffusion

Abstract In this chapter, application of the computational fluid dynamics tools to study thermodiffusion in fluid mixtures is presented. In addition to the description of the set of partial differential equations that are solved to study the transient behavior of the fluid components during the thermodiffusion process, two case studies are presented. The chapter also throws light on the significance of accurate calculations of local mixture density and viscosity values.

4.1 Introduction

In this chapter, we illustrate the use of computational fluid dynamics tools to study the transient separation behavior during the thermodiffusion process. CFD analysis is very helpful in understanding the evolution of the separation behavior and can help understand the experimental process better. Apart from the transient separation process in ideal thermodiffusive conditions, typical CFD studies in thermodiffusion include one or more of the following:

1. Understanding the effect of different levels of gravity acting on the fluid mixture enclosed in a domain that is subjected to a thermal gradient. Different levels of gravity are studied because experiments on earth experience a high gravitational force whereas on platforms like ISS the gravity levels are of $O(10^{-4})$ or $O(10^{-3})$. Even lower levels of disturbance are experienced by the experimental apparatus on free flying satellite platforms like FOTON, where the gravity level is of $O(10^{-6})$ or smaller.
2. As mentioned in Chap. 1, on platforms such as ISS and FOTON, there are disturbances emanating from various sources, resulting in undesirable vibrations that act as a source of convective force in the thermodiffusive process. Such a convective force can induce mixing of the components and will counter the separation behavior. CFD studies are often made to understand the effect of such vibrations that are recorded by the accelerometers on these space platforms [11].

S. Srinivasan and M.Z. Saghir, *Thermodiffusion in Multicomponent Mixtures*,
SpringerBriefs in Applied Sciences and Technology, DOI 10.1007/978-1-4614-5599-8_4,
© Springer Science+Business Media, LLC 2013

3. Using CFD studies one can study the effect of other disturbances during the course of the experiment such as an abrupt power failure, for instance [2]. Such failures will result in a heat loss through the boundaries, thereby adversely affecting the thermal boundary conditions that are to be maintained at constant values.
4. CFD studies can also be made to understand the impact of other inappropriate boundary conditions such as radiative heat loss and temperature fluctuations [2].

In the ensuing sections, following a description of a typical problem setup and the corresponding set of partial differential equations, known as the *governing equations*, to solve the problem setup, we will present two case studies, one each for a ternary and binary mixture. In these cases, the use of CFD simulations to understand the impact of vibrations on the concentration profile in the domain at steady state will be illustrated.

4.2 Problem Setup and Governing Equations

A schematic of a two-dimensional computational domain in which thermodiffusion is studied is shown in Fig. 4.1. The dimension of the domain is either square or rectangular, depending upon the experimental cell in which the experiment is conducted. While square cells are part of older experimental designs, most current experiments employ rectangular cells to minimize the end effects along the $x = 0$ and $x = L_x$ boundaries.

A temperature gradient is applied along the y-direction, as shown in this figure. Also, depending upon the environment, the cell can experience a *static* (fixed) or *variable* acceleration that are collectively represented by g_x and g_y in the two respective directions. Additionally, as indicated in this figure, the cell may be subjected to periodic oscillations in a direction that is orthogonal to the direction of the thermal gradient, and these are called *forced vibrations*.

Fig. 4.1 A schematic of the computational domain of dimension $L_x \times L_y$. In the CFD simulations, the various sources of convective disturbances such as g_x, g_y, and/or forced vibrations can be included to understand thermodiffusion in their presence

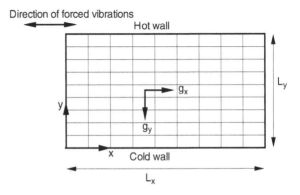

The governing equations to simulate this problem form a set of partial differential equations consisting of mass, species, momentum, and energy equations:

Mass conservation equation: For a weakly compressible flow in which the density fluctuations are a consequence of mild variations in the local temperature and concentration, the continuity equation is represented as:

$$\frac{\partial \rho}{\partial t} + \frac{\partial \rho u}{\partial x} + \frac{\partial \rho v}{\partial y} = 0, \tag{4.1}$$

where t is the time, u and v are the velocities in the x and y directions, respectively, and ρ is the density of the mixture. The mixture density can be computed using an appropriate equation of state that is coupled to this set of partial differential equations.

Species conservation equations: In addition to the mass conservation equation, species conservation equations are also included for the local concentration calculations. For the ternary mixtures, two species conservation equations are applied:

$$\left(\frac{\partial}{\partial t}(\rho c_1) + \frac{\partial}{\partial x}(\rho u c_1) + \frac{\partial}{\partial y}(\rho v c_1) \right) = \frac{\partial}{\partial x}\left(\rho \left(D_{11}\frac{\partial c_1}{\partial x} + D_{12}\frac{\partial c_2}{\partial x} + D_{T,1}\frac{\partial T}{\partial x} \right) \right)$$
$$+ \frac{\partial}{\partial y}\left(\rho \left(D_{11}\frac{\partial c_1}{\partial y} + D_{12}\frac{\partial c_2}{\partial y} + D_{T,1}\frac{\partial T}{\partial y} \right) \right), \tag{4.2}$$

$$\left(\frac{\partial}{\partial t}(\rho c_2) + \frac{\partial}{\partial x}(\rho u c_2) + \frac{\partial}{\partial y}(\rho v c_2) \right) = \frac{\partial}{\partial x}\left(\rho \left(D_{21}\frac{\partial c_1}{\partial x} + D_{22}\frac{\partial c_2}{\partial x} + D_{T,2}\frac{\partial T}{\partial x} \right) \right)$$
$$+ \frac{\partial}{\partial y}\left(\rho \left(D_{21}\frac{\partial c_1}{\partial y} + D_{22}\frac{\partial c_2}{\partial y} + D_{T,2}\frac{\partial T}{\partial y} \right) \right). \tag{4.3}$$

On the other hand, for binary mixtures, a single species equation is usually used:

$$\left(\frac{\partial}{\partial t}(\rho c_1) + \frac{\partial}{\partial x}(\rho u c_1) + \frac{\partial}{\partial y}(\rho v c_1) \right) = \frac{\partial}{\partial x}\left(\rho \left(D_{11}\frac{\partial c_1}{\partial x} + D_{T,1}\frac{\partial T}{\partial x} \right) \right). \tag{4.4}$$

In the above equations, c_i is the mole fraction of the ith species at the location (x,y) at time t. D_{ii} and D_{ij} represent the main term and the cross term diffusion coefficients that can be calculated using a molecular diffusion model [3, 4]. $D_{T,i}$ is the thermodiffusion coefficient of the ith species that can be calculated using a thermodiffusion model presented in Chap. 2.

It must be noted that the thermodiffusion as well as molecular diffusion coefficients depend on the mixture concentration that is nonuniform when $t > 0$ s, and constantly changing throughout the domain. As a result, it is more accurate to treat these coefficients as variables in computational domain, calculating them at each location at every time step. Nevertheless, some researchers simply use a fixed value of these coefficients in their calculations to minimize the computational time.

Finally, the species conservation at every location in the domain is completed with

$$\sum_i c_i = 1. \tag{4.5}$$

Momentum conservation equations: The x and y direction momentum conservation equations are:

$$\left(\frac{\partial}{\partial t}(\rho u) + \frac{\partial}{\partial x}(\rho uu) + \frac{\partial}{\partial y}(\rho vu) \right) = -\frac{\partial p}{\partial x} + \frac{\partial}{\partial x}\left(\eta \frac{\partial u}{\partial x} \right) + \frac{\partial}{\partial y}\left(\eta \frac{\partial u}{\partial y} \right) + \rho g_x, \tag{4.6}$$

$$\left(\frac{\partial}{\partial t}(\rho v) + \frac{\partial}{\partial x}(\rho uv) + \frac{\partial}{\partial y}(\rho vv) \right) = -\frac{\partial p}{\partial y} + \frac{\partial}{\partial x}\left(\eta \frac{\partial v}{\partial x} \right) + \frac{\partial}{\partial y}\left(\eta \frac{\partial v}{\partial y} \right) + \rho g_y. \tag{4.7}$$

In the above equations, p is the pressure and η is the dynamic viscosity. A good model for η of hydrocarbon mixtures is proposed by Lohrenz et al. [5]. For dilute mixtures, a weighted scheme can be used to determine the dynamic viscosity. g_x and g_y are the source terms that are discussed later.

Energy conservation equation: The energy equation to compute the temperature distribution in the domain is given in terms of the fluid temperature, T, as:

$$c_p \left(\frac{\partial}{\partial t}(\rho T) + \frac{\partial}{\partial x}(\rho uT) + \frac{\partial}{\partial y}(\rho vT) \right) = k \left(\frac{\partial^2 T}{\partial x^2} + \frac{\partial^2 T}{\partial y^2} \right), \tag{4.8}$$

where c_p and k are the mixture's specific heat capacity and the thermal conductivity, respectively. In this equation, it is assumed that the thermal conductivity of the mixture is constant and that there is no internal heat generation in the mixture.

The complete set of governing equations presented above, coupled with an appropriate equation of state and models for molecular and thermodiffusion coefficients, are simulated at every grid point in the computational domain to study the separation behavior over time. Usually, the duration of the simulation corresponds to the duration of the experiment.

4.2.1 Source Terms in Momentum Equation

The source terms in the momentum equations account for the natural gravitational force as well as the accelerations that are experienced by the system due to the vibrations of the platform on which the experimental cell is resting. For example, one can define the source term in the ith direction as

$$g_i = g_0 + A \sin(2\pi ft), \tag{4.9}$$

where g_0 is the static gravity, a constant acceleration that is always experienced by every point in the experimental cell. On earth, $g_0 = -9.8$ m s^{-2} and is present only in the y-direction. On space platforms that have reduced gravity environments, g_0 is a small nonzero value, and is present in g_x as well as g_y terms. The last term in (4.9) is a sinusoidal acceleration that is imposed on the system that has a frequency of f Hz and a magnitude of A m.

On space platforms like the ISS, the vibrations experienced by the experimental setup is recorded by on board accelerometers. These data can be included in the source term to simulate the actual environment. More precisely, for these simulations, g_x and g_y can be formulated as

$$g_x = g_x^{RMS},$$
$$g_y = g_y^{RMS}, \tag{4.10}$$

where

$$g_i^{RMS} = \sqrt{\frac{(g_i^{ISS}(1) - \bar{g}_i)^2 + (g_i^{ISS}(2) - \bar{g}_i)^2 + \cdots + (g_i^{ISS}(N) - \bar{g}_i)^2}{N}},$$
$$\bar{g}_i = \frac{\sum_{k=1}^{N} g_i^{ISS}(k)}{N}. \tag{4.11}$$

In the above equations, g_i^{ISS} is the raw acceleration in the ith direction that is recorded by the accelerometer every $1/N$ s, $\bar{g}_i$ is the average acceleration in a short (typically one second) interval, and g_i^{RMS} is the root mean square (RMS) acceleration in these intervals.

It must be noted that by using the RMS values, the acceleration is applied only along the positive side of each direction. This will result in a larger disagreements with the experiments. On the other hand, applying an average acceleration in the short intervals, one can simulate the positive and negative accelerations. This results in cancellation effects and a more closer agreement with the experimental results can be obtained. For this average representation, g_x and g_y can be formulated using the raw acceleration data as

$$g_i^{avg} = \frac{g_i^{ISS}(1) + g_i^{ISS}(2) + \cdots + g_i^{ISS}(N)}{N}. \tag{4.12}$$

Finally, for ease of computational implementation, one can study the raw accelerations in the Fourier domain, identify important frequencies that contribute to the accelerations, and construct a time varying source term like in (4.9). Usually, low frequencies (less than 10 Hz) are of interest since they have the most negative contribution to the thermodiffusive separation process [8].

4.3 Case Studies

The governing equations coupled with an appropriate equation of state and boundary conditions are solved for the unknowns, viz., c_i, T, p, u, v, and ρ, at every location in the domain. As mentioned before, the simulations are usually conducted for the duration of the experiment. The transient study allows us to understand the distribution of these parameters in the domain. In the following paragraphs, CFD investigations of a ternary and a binary mixture are presented in detail.

4.3.1 Case A: Methane–n-Butane–n-Dodecane Mixture

The experimental conditions, computational details, and the thermophysical properties of the ternary mixture for this case are summarized in Table 4.1. The mixture and the experimental conditions correspond to a recent experiment on the free flying satellite, FOTON-M3 [9]. The thermophysical properties have been obtained from the National Institute of Standards and Technology database [6].

The thermal gradient is applied by maintaining the hot wall at a temperature T_h and the cold wall at the colder temperature, T_c. Also, in the CFD simulations, a no slip boundary condition was applied. The entire setup is considered adiabatic with respect to the outer ambiance. To understand the diffusion process of all the components in the cavity, the complete set of governing equations presented in Sect. 4.2 must be solved.

For this case, two simulations are considered:

1. *Ideal*: In this simulation, the source terms (g_x and g_y) are set equal to zero, indicating an ideal zero-gravity environment.

Table 4.1 Experimental conditions, computational details, and thermophysical properties of nC_1–nC_4–nC_{12} mixture

Experimental conditions	
Mixt. concentration	0.2/0.5/0.3
Mixt. pressure, p [MPa]	35
Mixt. temp., T [K]	333
Hot wall temp., T_h [K]	338[a]
Cold wall temp., T_c [K]	328[b]
Computational domain	
Domain size, $L_x \times L_z$ [cm × cm]	4×0.6
Computational cells	60 × 9
Thermophysical properties	
Dynamic visc. [kg m^{-1}s^{-1}]	2.521×10^{-4}
Density [kg m^{-3}]	625.62
Thermal conductivity [W m^{-1}K^{-1}]	0.1054
Specific heat capacity [J kg^{-1}K^{-1}]	2399.0

[a] Boundary temperature applied along the right vertical wall
[b] Boundary temperature applied along the left vertical wall

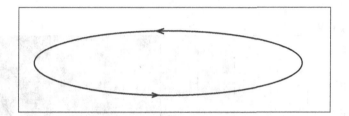

Fig. 4.2 The average flow profile of the nC_1–nC_4–nC_{12} mixture in the last 50 s in the domain after about 18.5 h

2. *ISS*: In this simulation, the demeaned RMS values of the raw accelerations recorded by the on board accelerometer on the ISS are applied as the source terms in the momentum equations.

It is found that in the *Ideal* environment the flow velocities are very weak with velocity magnitudes of the order of 1×10^{-50} m s^{-1}. In the *ISS* environment the velocities are higher with magnitudes of the order of 1×10^{-6} m s^{-1}. These higher velocities, which are primarily due to the imposed vibrations, result in a more structured flow with a well-established flow cell in the domain (c.f. Fig. 4.2).

The temperature distribution in the domain at the end of the simulations in the two environments is shown in Fig. 4.3. The vibrations on the ISS result in a marginal deviation of the temperature from an ideal linear profile, whereas much larger differences can be observed in the concentration distributions. This is shown for the two simulations in Fig. 4.4, where the distribution of nC_1 in the domain in the ideal as well as the ISS environment is presented. Clearly, the environment on ISS seems to hinder the thermodiffusion process. In fact the relative error for all three components along the center of the domain in the direction of the temperature gradient is as high as 50%. Such larger deviations are not seen in the temperature profile because the thermal boundary conditions are continuously imposed and they quickly correct most of these deviations from the ideal linear profile.

4.3.2   Case B: Water–Isopropanol Mixture

In this section, the CFD study for a binary mixture of water and isopropanol at atmospheric pressure and room temperature is presented. Table 4.2 summarizes the experimental conditions along with the computational details and the thermophysical properties of the mixture. Once again, two simulations pertaining to the ideal and the ISS environments can be compared.

While the Soret effect in this low pressure associating mixture causes isopropanol to separate toward the cold wall, the separation behavior is not the same in the two environments. The temperature and concentration of isopropanol in the domain

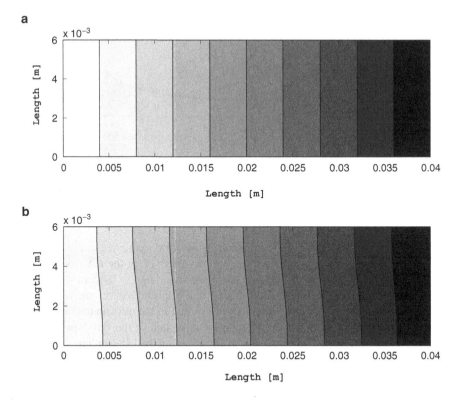

Fig. 4.3 Distribution of the temperature in the domain in the (**a**) ideal environment and (**b**) the ISS environment

at steady state in both environments are shown in Fig. 4.5. Once again, the ISS environment has a more disruptive effect on the thermodiffusion process.

The ISS environment results in the establishment of a single large vortex, similar to the one shown in Fig. 4.2. A large mixing is introduced in this environment and as a consequence, the separation of the components of the mixture is greatly subdued. In fact, the relative error in the concentration along the direction of the temperature gradient at the center of domain between the two simulations is close to 85%. Unlike the ternary n-alkane mixture just presented, the low pressure of this system means that the fluid flows more easily and hence this large error.

Note on the ISS vibrations: A special emphasis must be made regarding the fact that in the results shown in these figures, the ISS environment has been applied via the RMS expressions in (4.10) and (4.11), instead of the average value in (4.12). As mentioned earlier, if one uses the average values of the accelerations, then the simulation pertaining to the ISS environment will yield results that are much closer to the ideal environment.

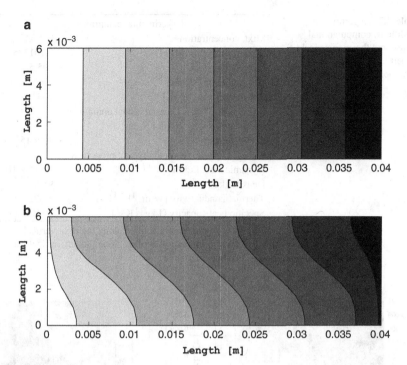

Fig. 4.4 Distribution of nC_1 in the domain in the (**a**) ideal environment and (**b**) the ISS environment

4.4   Relevance of Accurate Density and Viscosity Models

In applying CFD tools to study the transient flow and thermodiffusive behavior in the cavities, it is extremely important to use very accurate models, especially for density and viscosity calculations. If, for instance, one compares the CFD results from the application of two methods to calculate the local density of the mixture, viz., a weighted average scheme and *PC-SAFT* equation of state, to study the thermodiffusion in a water–isopropanol mixture that is subjected to forced vibrations, then it is found that the results are very different [7]. Likewise, induction of a viscosity model as opposed to constant viscosity assumption is also considered more accurate [10].

4.4.1   Density Calculations

In the simulations where the local mixture density is approximated using a simple weighted scheme of pure component densities, the values of ρ as well as their

Table 4.2 Experimental conditions, computational details, and thermophysical properties of water –isopropanol mixture

Experimental conditions	
Mixt. concentration	0.9/0.1
Mixt. pressure, p [MPa]	0.101325
Mixt. temp., T [K]	295.5
Hot wall temp., T_h [K]	298[a]
Cold wall temp., T_c [K]	293[b]
Computational domain	
Domain size, $L_x \times L_z$ [cm $\times$ cm]	1$\times$1
Computational cells	15 $\times$ 15
Thermophysical properties	
Dynamic visc. [kg m^{-1}s^{-1}]	1.036×10^{-3}
Density [kg m^{-3}]	993.36
Thermal conductivity [W m^{-1}K^{-1}]	0.522
Specific heat capacity [J kg^{-1}K^{-1}]	3991.0

[a] Boundary temperature applied along the right vertical wall
[b] Boundary temperature applied along the left vertical wall

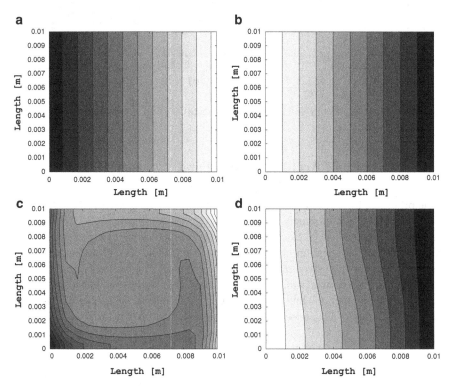

Fig. 4.5 Distribution of (**a**) concentration of isopropanol and (**b**) temperature in the domain in an ideal environment. Distribution of (**c**) concentration of isopropanol and (**d**) temperature in the domain in the ISS environment

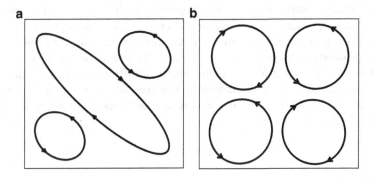

Fig. 4.6 CFD predictions of flow structures in the domain filled with water (90%) and iso-propanol (10%) via (**a**) calculation of the local mixture density using *PC-SAFT* equation of state and (**b**) the weighted average approximation of local mixture density

gradients are under-predicted in the domain [7]. On the contrary, *PC-SAFT* equation of state predicts a much larger value of density and its gradients [7]. These large density gradients magnify the instabilities in the flow and a flow profile of the type shown in Fig. 4.6a is obtained. With the weighted average scheme, the CFD simulations predict a weak flow with four vortices in the domain (c.f. Fig. 4.6b).

These calculations predict very different results for the thermodiffusion process. In the *PC-SAFT* case, the strong flow causes a complete mixing in the domain and the separation due to the Soret effect is minimal. On the other hand, unlike the *PC-SAFT* case, in the weighted-average scheme, the weak four-cell flow does not impede the thermodiffusion process and the CFD simulations predict a separation of isopropanol to the cold side [7].

4.4.2 Viscosity Calculations

In view of the small separations in the thermodiffusion process and computation-ally expensive simulations, it is an attractive option to simplify the problem by assuming a constant viscosity in the governing equations. However, a comparison of simulations using a constant viscosity assumption with the ones that employ a simple temperature-dependent linear model for viscosity shows that the latter predicts larger separations [10].

This trend, presented for ternary mixtures of nC_{12}–IBB–THN by Srinivasan and Saghir [10], is not surprising and can be attributed to the fact that as the temperature increases, viscosity decreases. This implies lower energy required to induce the molecular motion, i.e., lower activation energies. Thus, there is more pronounced molecular motion and it manifests itself in the form of larger separations, i.e., larger magnitude of the thermodiffusion parameters.

In summary, computational fluid dynamics is an invaluable tool to study the thermodiffusion process in fluid mixtures. In the case studies presented in this chapter, the ability to understand the distribution of relevant parameters in the domain at steady state has been demonstrated. By storing all the data of interest during the entire course of the simulation, such distributions can be studied at any t. One can also study the distribution at specific points in the domain and understand the evolution of the parameters such as temperature and concentration with time [1, 8].

References

1. Chacha M, Faruque D, Saghir MZ, Legros JC (2002) Solutal thermodiffusion in binary mixture in the presence of G-jitter. Int J Therm Sci 41:899–911. http://dx.doi.org/10.1016/S1290-0729(02)01382-0
2. Chacha M, Saghir MZ, Van Vaerenbergh V, Legros JC (2003) Influence of thermal boundary conditions on the double-diffusive process in a binary mixture. Philos Mag 83(17–18):2109–2129. http://dx.doi.org/10.1080/0141861031000108006
3. Ghorayeb K, Firoozabadi A (2000) Molecular, pressure, and thermal diffusion in non-ideal multicomponent mixtures. AIChE J 46(5):883–891. http://dx.doi.org/10.1002/aic.690460503
4. Leahy-Dios A, Firoozabadi A (2007) Unified model for nonideal multicomponent molecular diffusion coefficients. AIChE J 53(11):2932–2939
5. Lohrenz J, Bray BG, Clark C (1964) Calculating viscosities of reservoir fluids from their compositions. J Petrol Technol 16(10):1171–1176. http://dx.doi.org/10.2118/915-PA
6. NIST: Thermophysical properties of hydrocarbon mixtures database (2007) National Institute of Standards and Technology. Version 3.2
7. Parsa A, Srinivasan S, Saghir MZ (2012) Impact of density gradients on the fluid flow inside a vibrating cavity subjected to soret effect. Can J Chem Eng DOI: 10.1002/cjce.21666
8. Srinivasan S, Dejmeck M, Saghir MZ (2010) Thermo-solutal-diffusion in high pressure liquid mixtures in the presence of micro-vibrations. Int J Therm Sci 49:1613–1624
9. Srinivasan S, Saghir MZ (2009) Experimental data on thermodiffusion in ternary hydrocarbon mixtures. J Chem Phys 131:124,508
10. Srinivasan S, Saghir MZ (2010) Significance of equation of state and viscosity on the thermodiffusion coefficients of a ternary hydrocarbon mixture. High Temp High Pressur 39:65–81
11. Yan Y, Jules K, Saghir MZ (2007) A comparative study of G-jitter effect on thermal diffusion aboard the international spacestation. Fluid Dyn Mater Process J 3(3):231–245

Chapter 5
Application of Algebraic Correlations to Study Thermodiffusion

Abstract In Sect. 2.3 of Chap. 2, several simple algebraic expressions to predict the thermodiffusion process quantitatively in different types of mixtures were introduced. In this chapter, application of some of these algebraic equations for various liquid mixtures is presented.

5.1 Binary Liquid Mixtures

In this section we will evaluate some of the algebraic models from Sect. 2.3 of Chap. 2 to illustrate their performance in predicting the thermodiffusion coefficient in binary liquid mixtures. Several algebraic expressions are cast alongside each other to see a comparative performance of these expressions.

5.1.1 Binary n-Alkane Mixtures

We begin with an evaluation of the following five algebraic correlations that are applicable for binary n-alkane mixtures:

$$D_T = D_{T0}\delta M \left(1 + \lambda_1 \delta M\right), \tag{5.1a}$$

$$D_T = D_{T0}\delta M \left(1 + \lambda_1 \delta M + \lambda_2 (x_r - x_2)\right), \tag{5.1b}$$

$$D_T = D_{T0}\delta M \left(1 + \lambda_1 \delta M + \lambda_2 (x_r - x_2) + \lambda_3 (x_r - x_2)^2\right), \tag{5.1c}$$

$$D_T = K_0(x_r)(M_r - M_2)\frac{\beta}{\eta c_r c_2}, \tag{5.1d}$$

$$D_T = D_{T0} \left(x_r + \lambda_1 \Delta x + \lambda_2 (\Delta x)^2 + \lambda_3 \Delta M^{\text{mix}}\right) \frac{\beta \Delta M^{\text{mix}}}{\eta c_r c_2}, \tag{5.1e}$$

S. Srinivasan and M.Z. Saghir, *Thermodiffusion in Multicomponent Mixtures*,
SpringerBriefs in Applied Sciences and Technology, DOI 10.1007/978-1-4614-5599-8_5,
© Springer Science+Business Media, LLC 2013

Table 5.1 The second component and its mole fraction in the hexane−n-alkane mixture at 1 atm and 25°C

Component	nC_{10}	nC_{11}	nC_{12}	nC_{13}	nC_{14}	nC_{15}	nC_{16}	nC_{17}	nC_{18}
Mole fraction	0.337	0.355	0.336	0.319	0.303	0.289	0.276	0.264	0.253

The subscript indicates the number of carbon atoms in the n-alkane

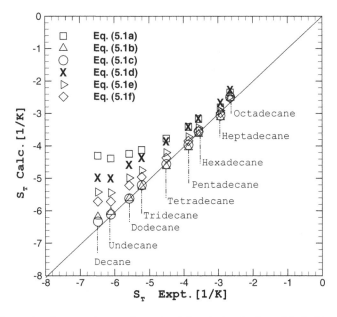

Fig. 5.1 Comparison between the experimental and the calculated values of S_T for binary mixtures of hexane with various higher alkanes. The experimental data for D_T and D are from [8]

$$D_T = D_{T0}\left(x_r + \lambda_1 \Delta x + \lambda_2 (\Delta x)^2 + \lambda_3 \Delta M\right)\frac{\beta \Delta M}{\eta c_r c_2}. \qquad (5.1f)$$

These expressions are (2.114), (2.119a), (2.119b), (2.120), (2.121), (2.122), respectively, in Chap. 2. We illustrate the use of these expressions to estimate the Soret coefficient in the mixtures of hexane and a higher n-alkane [8] at a temperature and pressure of 25°C and 1 atm. The composition of hexane in these mixtures is given in [8] and is summarized in Table 5.1. Since $S_T = D_T/D$, the experimental values of D have been used to translate the D_T values calculated by these correlations to the theoretical S_T in Fig. 5.1. The constants in these equations are summarized in Table 5.2, and the performance of these expressions are shown in Fig. 5.1. As seen in this figure, the largest deviation is in (5.1a) that neglects the impact of the composition of the mixture. On the other hand, this is taken into account in the other models that perform much better.

Table 5.2 Values of D_{T0} and λ_i in (5.1a)–(5.1f) for binary mixture of hexane with higher n-alkane mixtures

Equation #	D_{T0} [m²s⁻¹K⁻¹]	λ_1	λ_2	$\lambda_3 \times 10^4$ [mol g⁻¹]
(5.1a)	36.38×10^{-12}	1.36	–	–
(5.1b)	34.79×10^{-12}	1.317	0.29	–
(5.1c)	37.51×10^{-12}	1.393	−0.08	0.805
(5.1e)[a]	4.905×10^{-17}	−0.114	−0.122	−3.27
(5.1f)[a]	1.213×10^{-17}	−0.094	−1.053	−1.86

These constants are taken from [2] for (5.1a) and from [8] for (5.1b), (5.1c), (5.1e), and (5.1f)

[a] These constants are invariant for any binary n-alkane mixture

5.1.2 Binary Polymer Mixtures

We will now turn our attention to binary polymer mixtures. In particular, we will present the evaluation of some of the algebraic correlations with respect to dilute mixtures of polystyrene in several solvents. The experimental data for this is from the work of Hartung et al. [3], who have considered several mixtures at a temperature of 22°C. The different solvents considered are cyclooctane, cyclohexane, benzene, toluene, THF, ethyl acetate, and MEK.

The correlations that are evaluated include (2.123a), (2.126), (2.129), and (2.131) from Chap. 2 and are given as

$$D_T = \frac{\Delta_T}{\eta_{\text{eff}}}, \tag{5.2a}$$

$$D_T = \frac{D_b E_{A,s}}{RT^2}, \tag{5.2b}$$

$$D_T = \frac{8\beta_s \sqrt{A_p A_s} r_m^2}{27 v_s \eta_s} \left(1 - \sqrt{\frac{A_s}{A_p}}\right), \tag{5.2c}$$

$$D_T = \lambda \beta_s D_s^s. \tag{5.2d}$$

In (5.2a), $\Delta_T = 0.6 \times 10^{-14} \text{NK}^{-1}$ has been used [11]. In (5.2b), a hydrodynamic radius of $R_b = 0.44$ nm has been used to calculate D_b via (2.127). Further, the activation energy $E_{A,s}$ can be calculated using the solvent viscosity data as $E_{A,s} = -RT^2 d \ln \eta_s / dT$. In (5.2c), the radius of the monomer has been chosen as $r_m = 0.41$ nm. The Hamaker constants and the thermal expansion coefficients are taken from Hartung et al. [3]. Finally, in (5.2d), $\lambda = 1$ has been used and the self diffusion coefficient is from various resources as summarized in the work of Hartung et al. [3].

A comparative performance of all four correlations is plotted in Fig. 5.2. As seen in this figure, all correlations are fairly accurate in predicting the thermodiffusion quantitatively. In particular, it is interesting to note that thermodiffusion in polymers

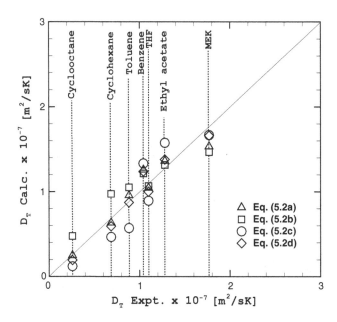

Fig. 5.2 Comparison between the experimental and the calculated values of D_T for dilute mixtures of polystyrene in various solvents. The experimental data for D_T are from Hartung et al. [3]. Figure modified from [3]

can be accurately quantified using just the solvent viscosity, as proposed in (5.2a). Of course, there is also a polymer-dependent parameter, Δ_T, in this relation that needs to be initially established via experimental data.

5.1.3 Binary Colloidal Mixtures

Temperature effects on the thermodiffusion process in colloidal mixtures has resulted in (2.132) of Chap. 2. In this section we will look at the application of this equation to colloidal mixtures. As per this equation, the Soret coefficient in several binary colloidal mixtures can be represented as a simple function of temperature as

$$S_T = S_T^\infty \left[1 - \exp\left(\frac{T^* - T}{T_0}\right) \right]. \tag{5.3}$$

As mentioned in Chap. 2, the three parameters, viz., S_T^∞, T^*, and T_0, in this equation can vary due to a change in the ionic strength of the solution. As a result of this, the Soret coefficients of the colloidal mixture can change. This is shown in Fig. 5.3 where (5.3) is plotted for the lysozome solutions with different concentration of NaCl as investigated by Iacopini and Piazza [4]. It is clearly

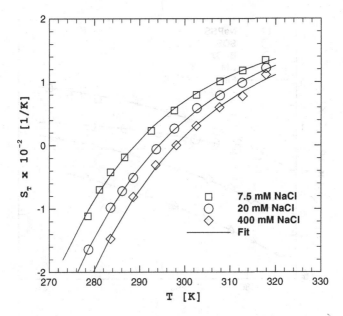

Fig. 5.3 Temperature dependence of S_T for lysozome solutions with different NaCl concentrations. The symbols are the experimental data from Iacopini and Piazza [4] for the experimental conditions therein. The *solid lines* are the fit using (5.3). Figure modified from [4]

evident that with an increase in the salt content, the mixture becomes increasingly thermophilic, i.e., S_T remains largely negative for a broad range of temperature. Put differently, there is an increasing tendency in the particles to move to the hot side.

Interestingly enough the above equation and the behavior is valid for several other systems. This is shown in Fig. 5.4 for aqueous solutions of NaPSS, SDS, β–lactoglobulin $-A$ (BLGA), and DM. While the latter two systems are thermophilic systems in which the particles move to the hot side, the first two systems tend to have a positive S_T for a wide temperature range, and would be called as the thermophobic mixtures. In these mixtures, the temperature dependence of S_T is primarily dictated by the thermal expansion coefficient of the solvent. A sign change in these binary mixtures usually occurs only when the operating temperature is lower than about 277 K.

Finally, if we consider the correlation

$$D_T = A\left(T^* - T\right), \tag{5.4}$$

an application to several aqueous systems shows a good prediction by this model. This linear dependence is seen in a wide variety of mixtures and is illustrated for a few in Fig. 5.5. It must be noted that the magnitude of A is independent of the size of the particles when the range of interactions between the particle and the solvent is short. This is based on the analysis of systems of polystyrene, SDS, and NaPSS [5].

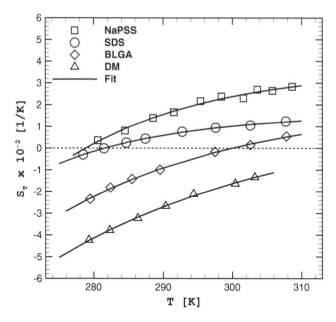

Fig. 5.4 Temperature dependence of S_T for various aqueous colloidal systems. The symbols are the experimental data in [5, 7] that also contain the experimental conditions. The *solid lines* are the fit using (5.3). Figure modified from [7]

Fig. 5.5 Temperature dependence of the thermodiffusion coefficient for different aqueous colloidal systems. The symbols are the experimental data from [5]. The *solid lines* are the linear fit using (5.4). Figure modified from [5]

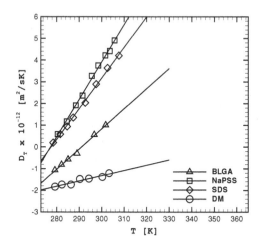

5.2 Ternary Mixtures

In ternary liquid mixtures too one can employ simpler algebraic expressions to quantify thermodiffusion. There are two approaches to this:

1. One can use the experimental thermodiffusion data from binary mixtures of the components in the ternary system. The experimental data of the binary mixtures can be combined algebraically to quantify thermodiffusion in a ternary mixture.
2. On the other hand, if the experimental data of the binary mixtures is unavailable, one can employ the correlations for the binary mixtures to calculate the thermodiffusion data. This calculated data can then be combined as above to quantify the thermodiffusion in a ternary mixture.

A simple expression for the thermodiffusion coefficient in a ternary mixture that makes use of the experimental value of $\mathscr{D}_T$ of the binary is [6]

$$\mathscr{D}_T^{(i)} = D_T^{(ij)} c_i c_j + D_T^{(ik)} c_i c_k, \tag{5.5}$$

where $\mathscr{D}_T^{(i)}$ is the ith thermodiffusion coefficient in the ternary mixture, and $D_T^{(ij)}$ is the thermodiffusion coefficient of the binary mixture of component i and j of the ternary mixture.

The second approach makes use of some physical properties of the binary mixtures in estimating the thermodiffusion coefficients in the ternary mixture as [1]

$$\mathscr{D}_T^{(i)} \frac{v^{(i)}}{\beta^{(i)}} = \frac{v^{(ij)}}{\beta^{(ij)}} D_T^{(ij)} c_i c_j + \frac{v^{(ik)}}{\beta^{(ik)}} D_T^{(ik)} c_i c_k, \tag{5.6}$$

where the superscript, (ij), in the kinematic viscosity (v) and the thermal expansion coefficient indicate that these values are for the binary mixture of components i and j in the ternary mixture.

A comparison of the above two correlations to calculate the thermodiffusion coefficients of THN and nC$_{12}$ in various ternary mixtures of nC$_{12}$–IBB–THN is shown in Figs. 5.6 and 5.7, respectively. The composition of these mixtures and the experimental conditions are discussed in detail by Blanco et al. [1]. While both correlations are good in predicting the sign as well as the magnitude of the thermodiffusion coefficient, the second correlation is more accurate than the first one. More precisely, while the first correlation has an average error of about 7%, the second one is within experimental error.

In the above calculations, the thermodiffusion coefficient of the binary mixture was obtained from the experiments. However, when experimental data are not available, we can calculate $D_T^{(ik)}$ using (2.116) as

$$D_T^{(ik)} = K(M_i - M_k) \frac{\beta^{(ik)}}{\eta^{(ik)} c_i c_k}. \tag{5.7}$$

The above results can be used in (5.5) and (5.6) as proposed in [9, 10]. For instance, in studying several ternary hydrocarbon mixtures of nC$_{12}$–IBB–THN, it has been found that for nC$_{12}$–IBB, IBB–THN and nC$_{12}$–THN binaries, the value of K is 2.88×10^{-14} m s^{-2}, 5.34×10^{-13} m s^{-2} and 6.0×10^{-14} m s^{-2}, respectively, in the above equation [9].

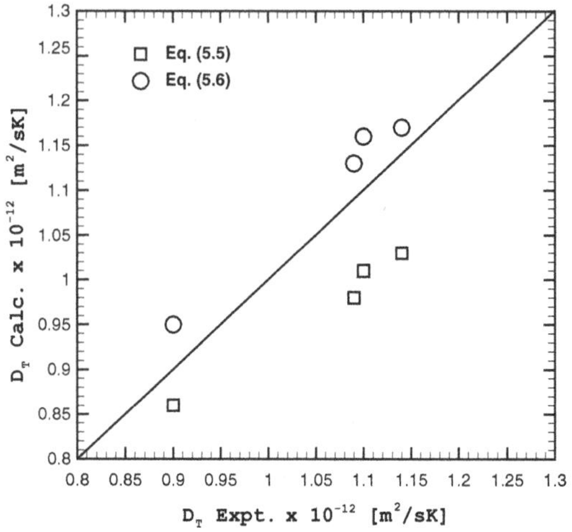

Fig. 5.6 Calculation of the
thermodiffusion coefficient of
THN in several ternary
mixtures of
$nC_{12}-IBB-THN$ using (5.5)
and (5.6). The experimental
conditions are in [1]

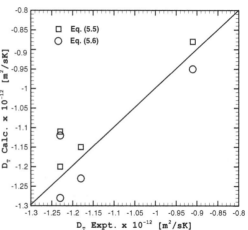

Fig. 5.7 Calculation of the
thermodiffusion coefficient of
nC_{12} in several ternary
mixtures of
$nC_{12}-IBB-THN$ using (5.5)
and (5.6). The experimental
conditions are in [1]

In summary, the application and the performance of various algebraic expressions introduced in Chap. 2 have been discussed in this chapter. It has been demonstrated that these simple equations can be easily employed to determine the separation behavior as well as the strength of the separation in different types of mixtures with little computational effort. There are several correlations for binary mixtures and this is a direct consequence of the exhaustive research being undertaken on binary systems. Also, extensive investigations on a variety of mixtures have resulted in numerous correlations for these systems.

On the other hand, in case of ternary liquid mixtures, there are only about two good correlations. Further, these are only valid with hydrocarbon mixtures and are yet to be evaluated with respect to other ternary mixtures. A lack of experimental data would mean that this evaluation will have to wait. In any case, all the presented correlations are fairly satisfactory in terms of their performance in the range of the parameters and/or the set of mixtures for which they are formulated. Nevertheless, further efforts are needed to develop a single universal correlation that is valid for any type of mixture and for a wide range of experimental conditions.

References

1. Blanco P, Bou-Ali M, Platten JK, de Mezquia DA, Madariaga JA, Santamaria C (2010) Thermodiffusion coefficients of binary and ternary hydrocarbon mixtures. J Chem Phys 132:114,506. http://dx.doi.org/10.1063/1.3354114
2. Blanco P, Bou-Ali M, Platten JK, Urteaga P, Madariaga JA, Santamaria C (2008) Determination of thermal diffusion coefficient in equimolar n-alkane mixtures: empirical correlations. J Chem Phys 129:174,504. http://dx.doi.org/10.1063/1.2945901
3. Hartung M, Rauch J, Köhler W (2006) Thermal diffusion of dilute polymer solutions: the role of solvent viscosity. J Chem Phys 125:214,904
4. Iacopini S, Piazza R (2003) Thermophoresis in protein solutions. Europhys Lett 63(2):247–253
5. Iacopini S, Rusconi R, Piazza R (2006) The "macromolecular tourist": universal temperature dependence of thermal diffusion in aqueous colloidal suspensions. Eur Phys J E 19:59–67
6. Larre JP, Platten JK, Chavepeyer G (1997) Soret effects in ternary systems heated from below. Int J Heat Mass Transfer 40:545–555
7. Piazza R, Parola R (2008) Thermophoresis in colloidal suspensions. J Phys Condens Matter 20:153,102
8. Srinivasan S, de Mezquia DA, Bou-Ali MM, Saghir MZ (2011) Thermodiffusion and molecular diffusion in binary n-alkane mixtures: experiments & numerical analysis. Philos Mag 91(34):4332–4344
9. Srinivasan S, Saghir MZ (2011) Thermodiffusion of ternary hydrocarbon mixtures: Part 1 – n-dodecane/isobutylbenzene/tetralin. J Non-Equilib Thermodyn 36:243–258
10. Srinivasan S, Saghir MZ (2012) Thermodiffusion of ternary hydrocarbon mixtures: Part 2 – n-decane/isobutylbenzene/tetralin. J Non-Equilib Thermodyn 37:99–113
11. Stadelmaier D, Köhler W (2009) Thermal diffusion of dilute polymer solutions: the role of chain flexibility and the effective segment size. Macromolecules 42:9147–9152

Chapter 6
Application of Neurocomputing Models to Study Thermodiffusion

Abstract The application of the principles of artificial neural networks, which were introduced in Chap. 2, to study thermodiffusion in binary liquids is discussed in this chapter. In this, the ability of this method to accurately evaluate the parameters relevant in thermodiffusion has been demonstrated for ionic and hydrocarbon mixtures. The method is compared with other approaches to highlight the accuracy of this approach in studying thermodiffusion.

6.1 Neural Network Studies in Thermodiffusion

A detailed formalism of artificial neural networks has been presented in Chap. 2. In this chapter, the application of the neurocomputing principles to study thermodiffusion in binary mixtures is taken up in detail. In particular, studies undertaken on two types of mixtures, viz., liquid metals and liquid hydrocarbons, are presented in the following sections.

6.1.1 Thermodiffusion in Binary Liquid Metals

Thermodiffusion is also called thermotransport in molten metals. In a recent investigation by Srinivasan and Saghir [17], thermotransport has been studied in binary liquid metals using the principles of artificial neural networks. A summary of the parameters corresponding to this study is given in Table 6.1 and are discussed below in detail.

6.1.1.1 Network Topology

Each neural network consisted of two layers, the first one with five nodes and a second one with one node. Thirty neural networks were generated and the average

S. Srinivasan and M.Z. Saghir, *Thermodiffusion in Multicomponent Mixtures*,
SpringerBriefs in Applied Sciences and Technology, DOI 10.1007/978-1-4614-5599-8_6,
© Springer Science+Business Media, LLC 2013

of the values predicted by these networks is taken as the prediction of the neural network. As discussed earlier, due to the randomness in the initiation of the network weights, different neural networks are likely to predict slightly different values of the outputs for the same inputs. The values of the various tolerance parameters for the termination of the training algorithm are as summarized in Table 6.1.

6.1.1.2 Database

The output of the neural network is the thermodiffusion factor, referred to as the thermotransport factor (α_T) in liquid metal literature. The inputs to the neural network include experimental parameters, thermophysical properties of the pure metals, and atomic structure and interaction parameters. The specific set of input parameters employed in [17] are

1. Mixture composition in mole fraction
2. Average mixture temperature
3. Temperature difference in the experiment
4. Molecular weight, density, viscosity, and thermal conductivity of the metals in the alloy
5. Pauling electron negativity of the metals in the alloy
6. Atomic radius of the metals in the alloy
7. Ionization energy of the metals in the alloy

A total of 140 experimental data points were included and the information of the input parameters were compiled from various sources in the literature. While 60% of this randomly shuffled, standardized, and normalized database was used for training, 10% of it was used for testing. The final 30% was used for the validation of the trained neural networks. As is customary in neural network studies, these three sets of data are mutually exclusive, i.e., a data point in one set is not found in the other.

6.1.1.3 Performance of Neural Network Model

The evaluation of the neural networks showed that the mean square error of the neural network prediction was about 0.5% for the evaluated experimental data. The results for some of the binary metal mixtures are also given in Fig. 6.1. From this figure, the following key observations can be made:

1. The neural network model can predict a wide range of the values of α_T, observed in the experiments.
2. It can predict the sign of α_T correctly, i.e., it can predict the direction of separation of the metals in the mixture.

Table 6.1 Summary of the artificial neural network model that was recently employed to study thermodiffusion in binary liquid metal mixtures [17]

Neural network details	
Number of neural networks	30
Topology	5-1[a]
Number of inputs	5
Number of outputs	1
Neural network training details	
tol_{mse}	2×10^{-5}
tol_t	200
tol_{grad}	1×10^{-9}
P_{max}	100,000
Database summary	
Size	140
Training %	60
Testing %	10
Validation %	30

[a] Five nodes in the first layer and one node in the output layer

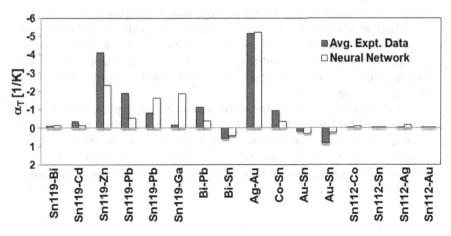

Fig. 6.1 Comparison of the experimental values of α_T of several binary molten metals with the ones calculated by the neural network. The experimental data are from [1, 9, 10, 18]

Apart from the above findings, a more thorough evaluation of the model with respect to the following parameters suggests a good performance of the neural network model:

1. *Sign change in isotopic mixtures:* The neural network model can correctly predict the sign change behavior noted in the experiments on isotopic mixtures of Tin. In particular, it is known from the experiments that the heavier isotopes move to the cold side and as the relative molecular weights of the two participating

Fig. 6.2 The contradicting trends of α_T versus temperature in two different molten metal systems. Figure modified from [17]

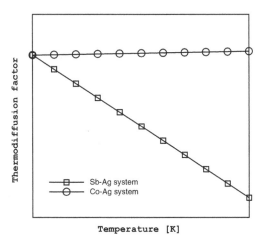

isotopes change from a negative to positive value, there is a sign change observed in the value of α_T. This experimentally observed trend is correctly predicted by the neural network.

2. *Effect of composition:* The neural network model can also be used to study the effect of composition in binary mixtures. Its predictive capabilities are highlighted by evaluating the effects of composition in a Na–K system. In a Na-rich Na–K system, the thermotransport factor is positive, whereas in a K-rich Na–K system, α_T is negative. This experimental result is accurately produced by the neural network model.

3. *Effect of temperature:* The ability of the neural network to predict the effects of temperature on the thermotransport factor has also been verified for at least two systems. In the Co–Ag system, α_T varies proportionately with the temperature, whereas in the Sb–Ag system, these two are inversely related (c.f. Fig. 6.2). The well-trained neural network model can predict these two contrasting trends accurately.

6.1.1.4   Thermal and Electronic Contributions

A significant advantage of modeling research over the experimental research is the former's ability to conduct detailed parametric investigation. With respect to the study of thermotransport in liquid metallic mixtures, the separation behavior in these liquids can be attributed to two types of contributions, viz., *thermal* and *electronic*. Thus, it is possible to study the two sets of contributions to understand the overall separation behavior. Specifically, by setting the weights of the input connections to the first layer (that correspond to the atomic structure and atomic interaction properties) to zero, we can obtain the thermal contribution to the thermotransport

Fig. 6.3 The effect of temperature of the system on the contribution of the electronic parameters toward the thermodiffusion in the Sb–Ag (*square* symbols) and Co–Ag (*circle* symbols) systems. Figure modified from [17]

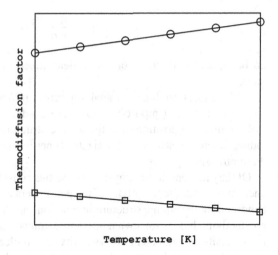

factor. Likewise by setting the weights of the input connections to the first layer that correspond to the thermophysical parameters to zero, one can understand the contribution of the electronic parameters.

For the Sb–Ag and Co–Ag systems, such an analysis shows that the thermal contribution to the thermotransport factor in these systems are constant at approximately -44 K^{-1} and -30 K^{-1}, respectively. The lower density and viscosity of Sb in the former system imply lower activation energy for the motion of the ions in the mixture, i.e., larger magnitude of α_T. On the other hand, higher density and viscosity of Co in the Co–Ag system imply larger activation energy requirements that results in the smaller magnitude of α_T.

While the thermal contributions are relatively constant over a range of temperatures, there is a linear relationship between the temperature and the electronic contribution (c.f. Fig. 6.3). Specifically, the Co–Ag system exhibits a positive slope whereas the Sb–Ag system exhibits a negative slope. These contradicting trends explain the opposite thermodiffusion behavior of the two systems in Fig. 6.2.

6.1.1.5 Sensitivity of Individual Parameters

In principle, it is logical to conduct a sensitivity analysis on each individual parameter to understand the effect/relevance of this parameter on the thermotransport process. This can be done by defining a measure of sensitivity of the ith parameter as

$$S_i = \frac{1}{n} \sum_{j=1}^{n} \Delta \alpha_T^{(j)}, \tag{6.1}$$

where n is the number of experiments recorded in the database. $\Delta \alpha_T$ gives a measure of the change affected by the parameter and is expressed as

$$\Delta \alpha_T = \frac{\alpha_T - \alpha_T'}{\alpha_T} \times 100, \tag{6.2}$$

α_T' being the thermotransport factor determined by switching off the parameter of interest.

With respect to binary metal mixtures, it has been found that the experimental conditions (imposed temperature gradient, average mixture temperature, and mixture composition) and the atomic structure/interaction parameters (atomic radius, electronegativity, and ionization energy) have a significant influence on the thermotransport process.

Of the different thermophysical properties considered, density has a strong influence, at par with the sensitivity of the parameters corresponding to the experimental conditions or the atomic structure/interaction parameters. On the other hand, thermal conductivity has a somewhat moderate impact on the separation process. The least sensitive parameters are viscosity and molecular weight. Intuitively, this is expected since the density is closely correlated with viscosity and molecular weight. (A component with higher molecular weight is expected to have higher density. Likewise, viscosity of a denser component is also usually higher). Such a strong correlation implies that the thermal contribution can be adequately represented using just the density and thermal conductivity of the pure components, whereas the impact of viscosity and molecular weight is implicitly included by using the pure metal density as an input to the neural network.

6.1.2   Thermodiffusion in Binary Liquid Hydrocarbon Mixtures

The neural network principles have also been successfully employed to study thermodiffusion in liquid hydrocarbon mixtures [16]. In the work of Srinivasan and Saghir [16], several series of binary n-alkanes were compiled into a database and subsequent to training a neural network on a part of this database, it was employed to study thermodiffusion trends in such mixtures. As in the binary molten metal mixtures, a summary of this neural network study is given in Table 6.2, and is discussed below in detail.

6.1.2.1   Network Topology

A schematic of a neural network for the hydrocarbon studies is shown in Fig. 6.4. As seen in this figure, each neural network consists of four layers, the number of nodes in these layers being five, four, one, and four, respectively. The last two layers have one node for each output parameter. The number of nodes in the second layer is the same as the number of nodes in the fourth layer, i.e., four.

Table 6.2 Summary of the artificial neural network model that was recently employed to study thermodiffusion in binary liquid hydrocarbon mixtures [16]

Neural network details	
Number of neural networks	50
Topology	5-4-1-4
Number of inputs	8
Number of outputs	5
Neural network training details	
tol_{mse}	2×10^{-3}
tol_t	200
tol_{grad}	1×10^{-9}
P_{max}	100,000
Database summary	
Size	145
Training %	70
Validation %	30

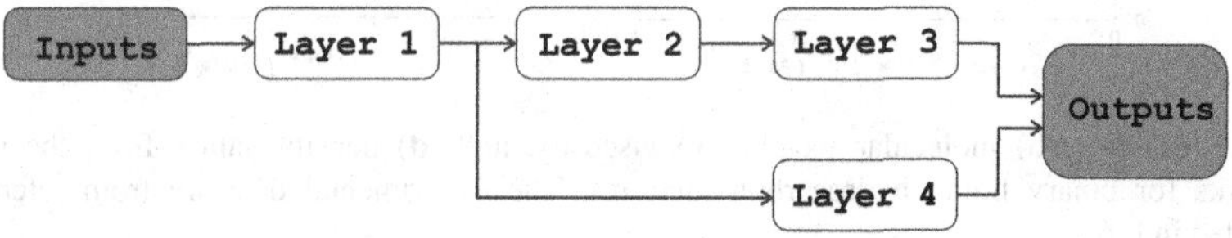

Fig. 6.4 A network topology to study thermodiffusion in binary liquid hydrocarbon mixtures

Again, it must be kept in mind that this is not the only topology. Other topologies can be considered as well, with varying number of network layers and nodes in each layer. Also, for the present case, only the number of nodes in the first layer is optimized, keeping the size of the second layer fixed.

Once again, an average of several neural networks represents a good estimate of the desired parameters. The typical parameters of the LMBP algorithm are summarized in Table 6.2.

6.1.2.2 Database

The inputs to the neural networks include the composition of the mixture and the pure component properties such as density, viscosity, and molecular weight. The output parameters that are to be predicted by the neural networks are thermodiffusion coefficient and mixture properties, viz., viscosity, density, thermal expansion coefficient, and molecular weight.

In the study in [16], a total of 145 experimental data points were included in the database and the information of the input parameters were compiled from various experiments reported in the literature. The data consisted of experiments

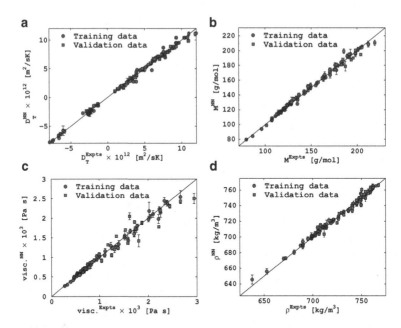

Fig. 6.5 (**a**) D_T, (**b**) molecular weight, (**c**) viscosity, and (**d**) density values from the neural networks for binary liquid hydrocarbon mixtures. The experimental data are from references compiled in [16]

of the following binary series: $nC_6–nC_i$, $nC_{10}–nC_i$, $nC_{12}–nC_i$, $nC_{15}–nC_i$, $nC_{16}–nC_i$ and $nC_{18}–nC_i$. While 70% of this randomly shuffled, standardized, and normalized database was used for training, the remaining 30% was used in the validation of the trained neural networks. Once again, these are mutually exclusive data sets, i.e., data point in one set is not found in the other.

6.1.2.3   Performance of the Neural Network Model

The ability of the neural network model to predict the experimental data of thermodiffusion coefficient and mixture's molecular weight, viscosity, and density are shown in Fig. 6.5a–d, respectively. As seen in these figures, the agreement with the experimental data is quite accurate for all the desired parameters. In particular, with respect to the thermodiffusion process it can be noted that a reasonably wide range of values of D_T, viz., large positive or negative D_T and small values of D_T in the neighborhood of zero, can be accurately predicted by a well-trained neural network.

Fig. 6.6 Prediction of the sign change in the thermodiffusion coefficients of nC_{12}–nC_i and nC_{10}–nC_i binary n-alkane mixtures. The experimental trend are based on the data from [2, 3, 8]

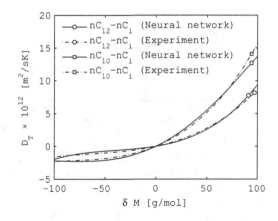

Sign Change in the Thermodiffusion Coefficient

Change in the preferred direction of separation is often observed in thermodiffusion experiments. The preferred direction of separation depends upon the difference between the participating molecules. In other words, depending upon the dissimilarity between the molecules, a change in the sign of D_T has been observed in the experiments.

The ability of a thermodiffusion model to predict this change in the sign is therefore an important test in establishing the reliability of the model. To this end, the neural network formalism when evaluated with respect to the experimental data of nC_{12}–nC_i and nC_{10}–nC_i binary n-alkane series shows an excellent predictive capability of the neurocomputing method. This is clearly evident in Fig. 6.6 where the sign change is not only accurately predicted but in fact D_T from the neural network is in close agreement with the experimental trend.

Parameter Analysis Using Neural Networks

The performance of the neural network, demonstrated in these figures, immediately points to the usability of the tool to study parameter effects in detail. To conduct such an analysis one parameter can be varied at a time in the normalized range of $[-1, 1]$, keeping the other parameters constant. Further, to understand the contribution of the parameter in the separation process, one can study the influence of the parameters on the mobility of the species. A quantitative measure of the relative ease of mobility of the species can be approximated using the mixture viscosity values as

$$\text{Mobility} = \frac{\eta^{(\text{mix,max})}}{\eta^{(\text{mix})}}. \tag{6.3}$$

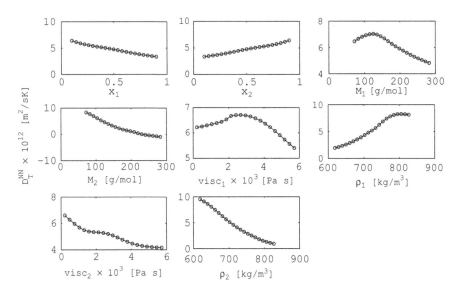

Fig. 6.7 Effect of the input parameters on D_T of the mixture of nC_{12} and nC_7 at 298 K and 1 atm

As an example, an analysis of the effects of parameters in a binary n-alkane mixture of nC_{12} and nC_7 at 298 K and 1 atm is shown in Fig. 6.7. The observed impact of each parameter on the separation process can be explained as follows:

Effect of x_1: Since nC_{12} has a higher viscosity, an increase in the concentration of this component (at the expense of the second component) increases the mixture viscosity. This decreases the mobility of the species in the mixture (c.f. (6.3)), ultimately decreasing the magnitude of D_T.

Effect of x_2: As expected, an opposite trend is observed when the composition of nC_7 is varied. With an increase in the concentration of this lower viscosity species, the mixture viscosity decreases and the overall mobility in the mixture increases. This results in larger D_T values.

Effect of M_1: While studying the effect of M_1, the molecular weight of the second species (M_2) is kept constant at 100.202 g mol^{-1}. Now, when normalized M_1 is varied in the range $[-1, 1]$, in the initial values of M_1, the mixture viscosity is lower and hence this disparity effect dominates and we observe a stronger thermodiffusion process, i.e., D_T increases. However, beyond a certain point, the mixture viscosity is too high, resulting in a lower mobility of the species in the mixture (c.f. (6.3)). As a result, the thermodiffusion process begins to get weaker, i.e., D_T decreases.

Of course, it must be noted that only one parameter is being varied at a time and so a nonzero D_T is observed even when normalized M_2 = normalized $M_1 = -1$ since the other pure component parameters are not the same.

Effect of M_2: As the molecular weight of the second component increases, not only does its disparity with the first component decrease, but there is also an increase in the mixture viscosity that impedes the mobility of the species. As a result, the thermodiffusive separation steadily decreases.

Effect of η_1: When normalized η_1 is varied in the range $[-1, 1]$, at the left end of this range, the overall mixture mobility is very high. In fact, for a small increase in η_1 there is a slight increase in the mobility because of disparity between the two mixture components. This enhances the thermodiffusion process. However, at higher values of η_1, there is a decrease in the mobility leading to a fast decrease in the value of D_T.

Effect of ρ_1: With the variation of the normalized density of the first component in the range $[-1, 1]$, the difference between the molecules increases, strengthening the thermodiffusion process. However, beyond approximately 770 kg m^{-3}, the mixture viscosity is high enough to diminish the mobility of the species. This has a negative effect on D_T.

Effect of η_2 and ρ_2: By increasing η_2 (ρ_2), the mixture property continuously approaches the value of η_1 (ρ_2) and this decreases the mixture disparity. Also, the effective mixture viscosity is continuously increasing, impeding the mobility of the mixture. This causes the thermodiffusion to decline rapidly, i.e., D_T continuously decreases.

6.2 Sensitivity and Diversity of the Parameters

In the selection of parameters that govern the input parameter set, it is advisable to conduct a prior sensitivity analysis to ensure that the output parameters are neither too sensitive nor too insensitive to the particular input parameter. A very sensitive parameter will result in a wild fluctuation in the parameter analysis graphs for instance. If this is the case, then it implies an over fitting of the experimental data set. On the other hand, if there are indications of acute insensitivity of the output parameter to the changes in input, then it is perhaps a sign that the input parameter does not have any impact on the output.

A quantitative estimate of the sensitivity of an output parameter, say D_T, to an input parameter can be given as

$$(\text{Sensitivity}) = \frac{1}{n} \sum_{j=1}^{n} \left| \frac{D_{T,\text{approx}}^{(j)} - D_{T,\text{true}}^{(j)}}{D_{T,\text{true}}^{(j)}} \right| \times 100, \qquad (6.4)$$

where n is the total number of data vectors in the database. To calculate $D_{T,\text{approx}}$ for a particular parameter, the link from the parameter to every node in the input layer is disabled. This can be done by simply setting the weights of these links to zero. $D_{T,\text{true}}$ is calculated without deactivating these links.

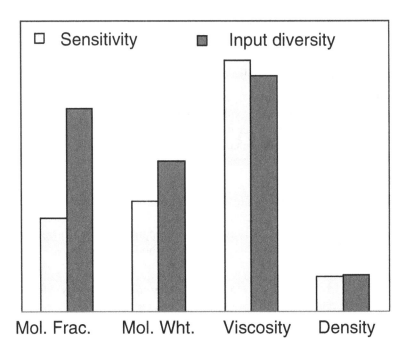

Fig. 6.8 Relative sensitivity of D_T to the input parameters and the diversity among the parameters

For the binary hydrocarbon mixtures presented in the previous section, the sensitivity analysis of the input parameters is shown in Fig. 6.8. For the binary mixtures, for each component in the mixture, there is an input value corresponding to the concentration, molecular weight, density, and viscosity. Hence, for each of these parameters, an average value of the sensitivity to the corresponding parameter for both components can be considered.

It must be noted that the ability of the neural network is dictated by the extent to which it is trained. This in turn is directly dependent on the diversity of the input parameter values in the database. Put differently, we can say that the sensitivity to a particular parameter is also governed by the diversity of the parameter in the input database. This is also evident in Fig. 6.8 in which a quantitative measure of the diversity of the ith parameter is obtained as

$$(\text{Range diversity})_i = \frac{p_i^{(\text{max})} - p_i^{(\text{min})}}{p_i^{(avg)}}. \tag{6.5}$$

In the above equation, the superscripts min, max, and avg represent the minimum, maximum, and average value of the parameter, p_i, in the database. Thus, a large sensitivity to viscosity but a relatively low sensitivity to density is attributed to the fact that the diversity of the density data in the database is less. On the contrary, there is a much wider range of viscosity data in the database.

6.3   Comparison with Other Approaches to Study Thermodiffusion

So far, we have looked at three different approaches to study thermodiffusion, viz., thermodynamic, algebraic, and neurocomputing. The underlying means and the purpose thereof in obtaining the thermodiffusion parameters from these methods are different and can be summarized as follows:

Thermodynamic models: Accurate values of D_T or α_T along with their signs have to be obtained by simulating the physics associated with the complex inter-particle interactions at the molecular or even atomic level in the thermodiffusion process.

Algebraic models: Accurate values of D_T or α_T along with their signs have to be obtained via simple algebraic correlations that have been developed after analyzing the data from several experiments. Such correlations are largely made up of simple physical parameters of the pure components or the mixture or both, and model constants, making them easy to evaluate to obtain a fairly accurate estimate of the thermodiffusion data.

Neurocomputing models: Accurate values of D_T or α_T along with their signs have to be obtained from the artificial neural networks that are trained and validated over an extensive set of experimental data. These trained networks are expected to capture the major trends in the experimental data set, enabling them to predict the thermodiffusion data in other mixtures.

To highlight the fact that the performance of the neural network approach, which is purely based on *data mining* strategies, is not inferior to the other two methods, this technique can be compared with the other approaches to study thermodiffusion. This is illustrated for the neural network models for thermodiffusion in molten metals as well as binary n-alkane mixtures.

6.3.1   Binary Molten Metal Mixtures

Over the years, different thermodynamic-based approaches to study thermodiffusion in liquid metals have been developed. In the literature of thermodiffusion, three important thermodynamic models are Haase [5], Kempers [6], and Shukla and Firoozabadi [11]. A relatively recent model to study thermodiffusion in molten metals is by Eslamian and Saghir [4], which accounts for the electronic contribution to the thermotransport in the molten metals. The expressions for α_T in these models are:

1. Haase [5]

$$\alpha_T = \frac{M_1 \bar{H}_2 - M_2 \bar{H}_1}{x_1 (M_1 x_1 + M_2 x_2)(\partial \mu_1 / \partial x_1)} \tag{6.6}$$

2. Kempers [6]

$$\alpha_T = \frac{\bar{V}_1 \bar{H}_2 - \bar{V}_2 \bar{H}_1}{x_1(\bar{V}_1 x_1 + \bar{V}_2 x_2)(\partial \mu_1 / \partial x_1)} \tag{6.7}$$

3. Shukla and Firoozabadi [11]

$$\alpha_T = \frac{\bar{V}_1 \bar{V}_2}{x_1 \bar{V}(\partial \mu_1 / \partial x_1)} \left[\frac{\bar{U}_1}{4\bar{V}_1} - \frac{\bar{U}_2}{4\bar{V}_2} \right] \tag{6.8}$$

4. Eslamian et al. [4]

$$\alpha_T = \frac{E_1^{\text{visc}} - E_2^{\text{visc}}}{x_1(\partial \mu_1 / \partial x_1)} + \frac{-|e|(zS_1 - z_1 S)TN}{x_1(\partial \mu_1 / \partial x_1)} \tag{6.9}$$

In the last equation, e, z, z_1, S, S_1, and N are electron charge, valence of the ions in the mixture, valence of the ions of component 1, thermoelectric power of the mixture, thermoelectric power of component 1, and Avogadro number, respectively. All other notations in these four equations are as explained in Chap. 2.

On comparing the neural network model with these models, it is found that the former is significantly better than the Firoozabadi, Hasse, and Kempers models. It performs at par or better with the Eslamian and Drickamer models. Overall, the predictions are in close agreement with the experimental data of α_T in several binary alloy mixtures. We arrive at this conclusion after studying the thermotransport factors in tin-based alloys and the bismuth-lead system [17].

6.3.2   Binary n-Alkane Mixtures

The neural network model for the hydrocarbon mixtures can also be evaluated with respect to the thermodynamic models in the literature. For illustration purpose we can choose the Shukla and Firoozabadi model. Additionally, to show a comparison with respect to the algebraic approach, the formulation of Madariaga et al. [7] can be evaluated. Recalling their expression, outlined in Chap. 2 (c.f. (2.120)), the thermodiffusion coefficient can be calculated algebraically as

$$D_T = k(x_j)(M_j - M_i)\frac{\beta}{\eta c_i c_j}, \tag{6.10}$$

where $k(x_j) = (5.34 x_j - 7 x_j^2 + 1.65 x_j^3) \times 10^{-14}$ m s^{-2}.

For the binary n-alkane series of nC$_6$–nC$_i$, a comparative plot of the neural network, algebraic, and thermodynamic formalisms is shown in conjunction with the trend in the experimental data in Fig. 6.9. It is clearly evident that the neural network strategy predicts the thermodiffusion coefficients fairly accurately and is in fact better than the other two approaches.

Fig. 6.9 D_T of binary n-alkane series of $nC_6–nC_i$ calculated using different approaches and compared with the trend in the experimental data of [15]

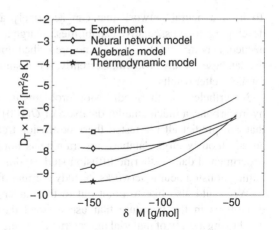

6.4 Concluding Remarks

6.4.1 Extension to Multicomponent Systems

Due to the simplicity of the implementation, the artificial neural network method can be used to study thermodiffusion in other types of binary mixtures. Additionally, while this methodology has been applied only to binary mixtures, it is not difficult to extend the model to ternary, four-component, or even multicomponent mixtures. However, this will require experimental data on a wide variety of mixtures. As such the number of experiments on even ternary mixtures are very limited. In any case, it must be noted that in complex mixtures of multicomponent systems, a much larger database with a lot of diversity in each input parameter in the database might be necessary to develop a well-trained network. In fact, given the complexity of the thermodiffusion process and the current state of the physics-based models, with adequate experimental data, neural network approach might be the most easily applicable method for multicomponent systems.

6.4.2 Small Database Size

An important point regarding the size of the database, training data set in particular, is pertinent here. In the literature, one can see that in several applications the size of the database for a NN study is a few orders of magnitude larger than the ones presented in this chapter. The size of the database in such studies is governed by several things:

1. A large range of the input parameters.
2. A variety of complicated trends in different ranges of the parameter.
3. Intricate interactions between several parameters.

To train a neural network that can accurately absorb all these trends thereby developing good associative capabilities, a large and diversified database might be needed. Having said this, there is no prescribed upper or lower limit on the size of the database for any problem. Currently, it is assumed that a large training database implies better results.

Nevertheless, in thermodiffusion problems in molten metals as well as liquid hydrocarbons, a much smaller database of $O(100)$ seems to yield neural networks that can predict all the tested thermodiffusion trends. Hence, one can argue that an enormously large database for training is not necessary. Given the dearth of experimental data in thermodiffusion studies, this serves as a boon to researchers aiming at using neural networks to study thermodiffusion processes.

We would also like to point out to the readers that there are studies in other disciplines in the literature that use a small database size [12–14]. In fact, in conducting a study of material microstructure using artificial neural networks, Singh and Bhadeshia [14] used a database with a size of just 15 experiments.

In summary, if one is looking for a universal tool to study thermodiffusion in a wide variety of liquid mixtures such as polymers, molten metals, and hydrocarbons, then the neurocomputing techniques are an attractive option. This novel engineering approach has been able to predict the separation behavior in binary liquid metals [17] as well as binary hydrocarbon mixtures [16]. The key findings in these works make an excellent case for pursuing this method and applying it to other mixtures of interest or even extending them to more complicated mixtures containing more components. This seemingly random and yet sophisticated method can predict the thermodiffusion data as well as important trends in the process by elegantly combining the principles of associative thinking with mathematical precision.

References

1. Bhat BN, Swalin RA (1972) Thermotransport of silver in liquid gold. Acta Metall Mater 20:1387–1396
2. Blanco P, Bou-Ali M, Platten JK, Urteaga P, Madariaga JA, Santamaria C (2008) Determination of thermal diffusion coefficient in equimolar n-alkane mixtures: empirical correlations. J Chem Phys 129:174,504. http://dx.doi.org/10.1063/1.2945901
3. Blanco P, Polyakov P, Bou-Ali M, Wiegand S (2008) Thermal diffusion and molecular diffusion values for some alkane mixtures: a comparison between thermogravitational column and thermal diffusion forced rayleigh scattering. J Phys Chem 112(28):8340–8345. http://dx.doi.org/10.1021/jp801894b
4. Eslamian M, Sabzi F, Saghir MZ (2010) Modeling of thermodiffusion in liquid metal alloys. Phys Chem Chem Phys 12:13,835–13,848
5. Haase R (1969) Thermodynamics of irreversible processes. Addison-Wesley, Reading
6. Kempers LJTM (2001) A comprehensive thermodynamic theory of the soret effect in a multicomponent gas, liquid, or solid. J Chem Phys 115:6330–6341
7. Madariaga JA, Santamaria C, Bou-Ali M, Urteaga P, De Mezquia DA (2010) Measurement of thermodiffusion coefficient in n-alkane binary mixtures: composition dependence. J Phys Chem B 114:6937–6942. http://dx.doi.org/10.1021/jp910823c

8. Perronace A, Leppla C, Leroy F, Rousseau B, Wiegand S (2002) Soret and mass diffusion measurements and molecular dynamics simulations of n-pentane-n-decane mixtures. J Chem Phys 116:3718

9. Praizey JP (1989) Benefits of microgravity for measuring thermotransport coefficients in liquid metallic alloys. Int J Heat Mass Transfer 32(12):2385–2401. http://dx.doi.org/10.1016/0017-9310(89)90199-3

10. Praizey JP, Van Vaerenbergh S, Garandet JP (1995) Thermomigration experiment on board EURECA. Adv Space Res 16(7):205–214. http://dx.doi.org/10.1016/0273-1177(95)00161-7

11. Shukla K, Firoozabadi A (2000) Theoretical model of thermal diffusion factors in multicomponent mixtures. AIChE J 46:892–900

12. Sidhu G, Bhole SD, Chen DL, Essadiqi E (2011) Determination of volume fraction of bainite in low carbon steels using artificial neural networks. Comput Mater Sci 50(12):3377–3384

13. Sidhu G, Bhole SD, Chen DL, Essadiqi E (2012) Development and experimental validation of a neural network model for prediction and analysis of the strength of bainitic steels. Mater Design 41:99–107

14. Singh SB, Bhadeshia HKDH (1998) Estimation of bainite plate-thickness in low-alloy steels. Mater Sci Eng A 245(1):72–79

15. Srinivasan S, de Mezquia DA, Bou-Ali MM, Saghir MZ (2011) Thermodiffusion and molecular diffusion in binary n-alkane mixtures: experiments & numerical analysis. Philos Mag 91(34):4332–4344

16. Srinivasan S, Saghir MZ (2012a) A neurocomputing model to calculate the thermo-solutal diffusion in liquid hydrocarbon mixtures. Neural Comput Appl. DOI: 10.1007/s00521-012-1217-6

17. Srinivasan S, Saghir MZ (2012b) Modeling of thermotransport phenomenon in metal alloys using artificial neural networks. Appl Math Modell. DOI:10.1016/j.apm.2012.06.018

18. Winter FR, Drickamer HG (1955) Thermal diffusion in liquid metals. J Phys Chem 59(12):1229–1230

Index

S. Srinivasan and M.Z. Saghir, *Thermodiffusion in Multicomponent Mixtures*,
SpringerBriefs in Applied Sciences and Technology, DOI 10.1007/978-1-4614-5599-8,
© Springer Science+Business Media, LLC 2013